The first two authors were partially supported by the National Science Foundation and the Office of Naval Research.

The production of this volume was supported in part by the Fonds pour la Formation de Chercheurs et l'Aide à la Recherche (Fonds FCAR) and the Natural Sciences and Engineering Research Council of Canada (NSERC).

2000 *Mathematics Subject Classification.* Primary 60–XX; Secondary 62–XX, 76–XX, 82–XX, 86–XX, 94–XX.

For additional information and updates on this book, visit
www.ams.org/bookpages/crmm-25

Library of Congress Cataloging-in-Publication Data

Majda, Andrew, 1949–
Information theory and stochastics for multiscale nonlinear systems / Andrew J. Majda, Rafail V. Abramov, Marcus J. Grote.
p. cm. — (CRM monograph series ; v. 25)
Includes bibliographical references.
ISBN 0-8218-3843-1 (alk. paper)
1. Information theory in mathematics. 2. Stochastic analysis. 3. Nonlinear systems. I. Abramov, Rafail V., 1978– II. Grote, Marcus J. III. Title. IV. Series.

QA10.4.M35 2005
003′.75—dc22 2005050718

∞ The paper used in this book is acid-free and falls within the guidelines
established to ensure permanence and durability.
This volume was submitted to the American Mathematical Society
in camera ready form by the Centre de Recherches Mathématiques.
Visit the AMS home page at http://www.ams.org/

10 9 8 7 6 5 4 3 2 1 10 09 08 07 06 05

Contents

Overview on Multiscale Modeling in Complex Nonlinear Systems

It is clear to the practitioners in many disciplines that major developments in the future involve multiscale modeling for complex nonlinear systems including their nonlinear statistical features, novel computational strategies to assess interactions across scales, and reliable ensemble predictions from given data. Such strategies are emerging in diverse disciplines ranging from nanotechnology, material science, and biomolecular dynamics to the coupled atmosphere-ocean system. Novel strategies to deal with these extremely difficult issues are being developed at a rapid rate by applied mathematicians in a variety of contexts [**4, 39, 13, 22, 36, 38**]. The article by Weinan E and Engquist [**39**] contains an extensive bibliography of these developments in scientific disciplines outside atmosphere/ocean science. The present book emphasizes some of these novel developments in atmosphere/ocean science.

One of the grand challenges of contemporary science is a comprehensive predictive model for the atmosphere and coupled climate system. This is one of the most difficult multiscale problems in contemporary science because there is an incredible range of strongly interacting anisotropic nonlinear processes over many spatio-temporal scales; contemporary comprehensive computer models, GCM's, are currently incapable of adequately resolving or parameterizing many of these interactions on time scales appropriate for seasonal prediction as well as climate change projections. An overview for mathematicians can be found in the recent article by the first author [**22**].

Thus, the multiscale problems in atmosphere ocean dynamics serve as important prototype problems for developing new systematic multiscale strategies which are valuable in other scientific disciplines ranging from nanotechnology to macromolecular dynamics to protein folding, etc. The societal impacts for these efforts are also large; it has been estimated recently that a one month increase in lead time for El Niño prediction would save 100 billion dollars for the world community.

Basic questions which drive climate research are the prediction of the weather from 1 to 14 days, the prediction of climate variations on seasonal to yearly time scales and finally, climate change projections on decadal and centennial time scales as well as quantifying the uncertainty associated with these predictions.

One of the striking recent observational discoveries is the profound impact of variations in the tropics on all of these problems. The primary issue in the influence of the tropics occurs through the interaction and organization of clouds into clusters, super clusters, and planetary scale dynamics, an inherently fully nonlinear multiscale process. For climate change, water vapor is the most important greenhouse gas and the microphysical processes in clouds are a key mechanism for radiative feedback. In fact, only a 4% change in average cloudiness would overwhelm the effects of CO_2 in climate change. Current evidence suggests that a few global planetary teleconnection patterns, such as the Pacific/North American pattern, often summarize the weather and climate impact of the tropics for the mid-latitude atmosphere. Since it will be impossible to run resolved coupled atmosphere/ocean comprehensive numerical models for the atmosphere for climate change projections, reduced models involving these basic large scale patterns are of central importance.

Emerging mathematical and computational strategies for multiscale modeling in atmosphere/ocean science. A new perspective on several of the issues discussed above for climate dynamics has been developed through the paradigm of modern applied mathematics, where rigorous and asymptotic multiscale mathematical theory, the development of prototype model problems and novel computational strategies all interact simultaneously in understanding these complex scientific problems. These emerging mathematical/computational strategies include the following:

1) Systematic multiscale asymptotic modeling for the tropics [**3, 23, 25, 24, 28**]
2) Novel stochastic models for unresolved features of tropical convection [**17, 26**]
3) Systematic mathematical strategies for low-dimensional stochastic mode reduction in climate [**5, 8, 33, 29, 30, 31, 32**]
4) Quantifying uncertainty in ensemble predictions and loss of information in coarse-grained stochastic models through information theory [**1, 2, 10, 14, 18, 19, 27**]

Through the serendipity of modern applied mathematics, the issues in 2) have driven new systematic mathematical strategies to coarse-grain stochastic lattice models with adsorption/desorption and diffusion coupled to continuum problems at mesoscopic scales; the result is a new class of coarse grained Monte Carlo methods for material science with simultaneously a speed up of order a billion times in computational labor while retaining fidelity with key features of the microscopic fluctuations [**12, 13, 15**]. In another different multi-disciplinary application, the methodology in 3) has been applied to stochastic immersed boundary numerical methods for microfluid particle suspensions, as, for example, molecular motors in cellular biology [**20, 21**].

Systematic applications of ideas from equilibrium statistical mechanics lead to a different class of novel promising strategies for assessing the unresolved scales of motion in many problems in science and engineering. Geophysical flows provide an important prototype context for developing new methods for underresolved models in science and engineering. The wide range of unresolved scales of motion with highly inhomogeneous behavior for the atmosphere and ocean on Earth, as well as for other planets like Jupiter, require novel statistical strategies to assess the nontrivial impact of these unresolved features on the larger scales. Novel systematic applications of ideas from equilibrium statistical mechanics to these inhomogeneous geophysical flows have lead to promising new statistical strategies for the ocean [**6, 7, 11, 16**], the atmosphere [**9, 35**], and the giant planets such as Jupiter [**37**] in agreement with contemporary observations.

In the remaining chapters of this book we show how mathematical ideas from information theory and stochastic analysis lead to novel ways to understand and assess the behavior of complex multiscale nonlinear systems. The interested reader can see the paradigm of modern applied mathematics, mentioned above, applied to diverse problems; in particular, the design and use of appropriate simplified mathematical models which yield special insight into much more complex real world phenomena and applications. The material developed below is loosely based on A. Majda's Aisenstadt Lectures at the Center for Research in Mathematics at the University of Montreal, November 8-12, 2004. There are three chapters below which, by design, can be read essentially independently of each other. There is new research material, presented for the first time here, interspersed throughout Chapters 2 and 3. The material in Chapter 1 is a summary of ideas where many omitted details of the research program can be found in the recent book by Majda and Wang [**34**]. The first author would like to thank his collaborators and friends, Eric Vanden-Eijnden, Bruce Turkington, Richard Kleeman, Ilya Timofeyev, Xiaoming Wang, Daan Crommelin and Christian Franzke for their explicit and implicit contributions to the material presented below. The first author also thanks the head of the special year on multiscale problems Prof. Anne Bourlioux of the University of Montreal for her kindness and help throughout the lecture series as well as with the follow-up manuscript developed here. The first two authors acknowledge generous support from the National Science Foundation and Office of Naval Research during the preparation of this material.

References

1. R. Abramov and A. J. Majda, *Quantifying uncertainty for non-Gaussian ensembles in complex systems*, SIAM J. Sci. Comput. **26** (2004), no. 2, 411–447.
2. R. Abramov, A. J. Majda, and R Kleeman, *Information theory and predictability for low frequency variability*, J. Atmospheric Sci. (2004), in press.
3. J. Biello and A. J. Majda, *A new multiscale model for the Madden-Julian Oscillation*, J. Atmospheric Sci. (2004), in press.
4. A. Brandt, *Multigrid methods in lattice field computations*, Nuclear Physics B Proc. Suppl. **26** (1992), 137–180.
5. D. Crommelin and A. J. Majda, *Strategies for model reduction: Comparing different optimal bases*, J. Atmospheric Sci. **61** (2004), no. 17, 2206–2217.
6. M. DiBattista and A. J. Majda, *An equilibrium statistical theory for large-scale features of open-ocean convection*, J. Phys. Oceanogr. **30** (2000), 1325–1353.
7. M. DiBattista, A. J. Majda, and J. Marshall, *A statistical theory for the "patchiness" of open-ocean deep convection: The effect of preconditioning*, J. Phys. Oceanogr. **32** (2002), 599–626.
8. C. Franzke, A. J. Majda, and E. Vanden-Eijnden, *Low-order stochastic mode reduction for a realistic barotropic model climate*, J. Atmospheric Sci. (2004), in press.
9. J. Frederiksen, M. Dix, and S. Kepert, *Systematic energy errors and the tendency toward canonical equilibrium in atmospheric circulation models*, J. Atmospheric Sci. **53** (1996), 887–904.
10. K. Haven, A. J. Majda, and R. Abramov, *Quantifying predictability through information theory: Small sample estimation in a non-Gaussian framework*, J. Comput. Phys. (2004), in press.
11. G. Holloway, *Representing topographic stress for large-scale ocean models*, J. Phys. Oceanogr. **22** (1992), 1033–1046.
12. M. Katsoulakis, A. J. Majda, and D. Vlachos, *Coarse-grained stochastic processes and Monte-Carlo simulations in lattice systems*, J. Comput. Phys. **186** (2003), no. 1, 250–278.
13. ———, *Coarse-grained stochastic processes for lattice systems*, Proc. Nat. Acad. Sci. **100** (2003), 782–787.
14. M. Katsoulakis and J. Trashorras, *Information loss in coarse-graining of stochastic particle dynamics*, J. Stat. Phys. (2004), in press.
15. M. Katsoulakis and D. Vlachos, *Coarse-grained stochastic processes and kinetic Monte-Carlo simulators for the diffusion of interacting particles*, J. Chem. Phys **199** (2003), no. 18, 9412–9427.

16. E. Kazantsev, J. Sommeria, and J. Verron, *Subgrid-scale Eddy parameterization by statistical mechanics in a barotropic ocean model*, J. Phys. Oceanogr. **28** (1998), 1017–1042.
17. B. Khouider, A. J. Majda, and M. Katsoulakis, *Coarse-grained stochastic models for tropical convection and climate*, Proc. Nat. Acad. Sci. U.S.A. **100** (2003), 11941–11946.
18. R. Kleeman, *Measuring dynamical prediction utility using relative entropy*, J. Atmospheric Sci. **59** (2002), 2057–2072.
19. R. Kleeman, A. J. Majda, and I. Timofeyev, *Quantifying predictability in a model with statistical features of the atmosphere*, Proc. Nat. Acad. Sci. **99** (2002), 15291–15296.
20. P. Kramer and A. J. Majda, *Stochastic mode reduction for particle-based simulation methods for complex microfluid systems*, SIAM J. Appl. Math. **64** (2003), no. 2, 401–422.
21. ———, *Stochastic mode reduction for the immersed boundary method*, SIAM J. Appl. Math. **64** (2003), no. 2, 369–400.
22. A. J. Majda, *Real world turbulence and modern applied mathematics*, Amer. Math. Soc., Providence, RI, 2000.
23. A. J. Majda and A. Bertozzi, *Introduction to PDEs and waves for the atmosphere and ocean*, Amer. Math. Soc., Providence, RI, 2003.
24. A. J. Majda and J. Biello, *The nonlinear interaction of barotropic and equatorial baroclinic Rossby waves*, J. Atmospheric Sci. **60** (2003), no. 15, 1809–1821.
25. ———, *A multiscale model for tropical intraseasonal oscillations*, Proc. Nat. Acad. Sci. **101** (2004), 4736–4741.
26. A. J. Majda and B. Khouider, *Stochastic and mesoscopic models for tropical convection*, Proc. Nat. Acad. Sci. **99** (2002), 1123–1128.
27. A. J. Majda, R. Kleeman, and D. Cai, *A framework for predictability through relative entropy*, Methods Appl. Anal. **9** (2002), 425–444.
28. A. J. Majda and R. Klein, *Systematic multiscale models for the tropics*, J. Atmospheric Sci. **60** (2003), no. 2, 393–408.
29. A. J. Majda, I. Timofeyev, and E. Vanden-Eijnden, *Models for stochastic climate prediction*, Proc. Nat. Acad. Sci. **96** (1999), 14687–14691.
30. ———, *A mathematical framework for stochastic climate models*, Comm. Pure Appl. Math. **54** (2001), no. 8, 891–974.
31. ———, *A priori tests of a stochastic mode reduction strategy*, Phys. D **170** (2002), no. 3–4, 206–252.
32. ———, *Stochastic mode reduction in climate*, J. Atmospheric Sci. **60** (2003), no. 14, 1705–1722.
33. ———, *Low-dimensional chaotic dynamics versus intrinsic stochastic noise: A paradigm model*, Phys. D **199** (2004), no. 3–4, 339–368.
34. A. J. Majda and X. Wang, *Nonlinear dynamics and statistical theories for basic geophysical flows*, Cambridge University Press, New York, 2005.
35. W. Schubert, M. Montgomery, R. Taft, T. Guinn, S. Fulton, J. Kossin, and J. Edwards, *Polygonal eyewalls, asymmetric eye contraction, and potential vorticity mixing in hurricanes*, J. Atmospheric Sci. **56** (1999), 1197–1223.
36. C. Theodoropoulos, Y.-H. Qian, and I. Kevrekidis, *"Coarse" stability and bifurcation analysis using time-steppers: A reaction-diffusion example*, Proc. Nat. Acad. Sci. **97** (2000), no. 18, 9840–9843.
37. B. Turkington, A. J. Majda, K. Haven, and M. DiBattista, *Statistical equilibrium predictions of jets and spots on Jupiter*, Proc. Nat. Acad. Sci. **98** (2001), 12346–12350.
38. E. Vanden-Eijnden, *Numerical techniques for multiscale dynamical systems with stochastic effects*, Commun. Math. Sci. **2** (2003), no. 2, 385–391.
39. E. Weinan and E. Bjorn, *The heterogeneous multiscale methods*, Commun. Math. Sci. **1** (2003), no. 1, 87–133.

CHAPTER 1

Information Theory, Predictability, Jupiter's Great Red Spot, and Equilibrium Statistical Mechanics

1.1. Introduction to information theory

A continuous probability density, $p(\lambda)$, on the line satisfies the two requirements

$$p(\lambda) \geq 0, \quad \int_{\mathbb{R}} p(\lambda)\, d\lambda = 1. \tag{1.1}$$

Such probability densities arise as the probability distribution function of some random variable, q, so that

$$\text{Prob}\{\alpha \leq q \leq \beta\} = \int_{\alpha}^{\beta} p(\lambda)\, d\lambda. \tag{1.2}$$

The book of Lamperti [**24**] is an elementary reference for the material presented next.

Let $\langle\cdot\rangle$ denote the expected value of a random variable over the probability space where the random variable, q, is defined. Then, with the density $p(\lambda)$ in (1.1),

$$\langle F(q)\rangle = \int_{\mathbb{R}} F(\lambda)p(\lambda)\, d\lambda. \tag{1.3}$$

The mean, $\bar{q}$, and variance, σ^2, of a random variable are defined respectively by

$$\begin{aligned} \bar{q} &\equiv \langle q\rangle = \int_{\mathbb{R}} \lambda p(\lambda)\, d\lambda, \\ \sigma^2 &\equiv \langle (q-\bar{q})^2\rangle = \int_{\mathbb{R}} (\lambda - \bar{q})^2 p(\lambda)\, d\lambda. \end{aligned} \tag{1.4}$$

Perhaps the most well-known and important examples are Gaussian densities which are uniquely determined by their mean and variance as follows: A Gaussian distribution with given mean $\bar{\lambda}$ and variance σ has the probability density function

$$p_{\bar{\lambda},\sigma}(\lambda) = \frac{1}{\sqrt{2\pi}\sigma} e^{-(\lambda-\bar{\lambda})^2/2\sigma^2} \tag{1.5}$$

so that

$$\bar{\lambda} = \int \lambda p_{\bar{\lambda},\sigma}(\lambda)\, d\lambda, \quad \sigma^2 = \int (\lambda - \bar{\lambda})^2 p_{\bar{\lambda},\sigma}(\lambda)\, d\lambda. \tag{1.6}$$

Below we often use the related notation, $p(\vec{\lambda})$, for a probability density in $\mathbb{R}^N$ and say $p \in PM$.

Information theory was developed by Shannon [**41**] in pioneering work; the ideas presented throughout this chapter are strongly influenced by the seminal article of

Jaynes [**17**]. One outgrowth of this work is that the entropy, $S(p)$, naturally arises axiomatically as the measure of the absolute lack of information in a probability density $p(\lambda)$ with

$$S(p) = -\int_{\mathbb{R}^N} p \ln p \, d\vec{\lambda}. \tag{1.7}$$

Similarly, the relative entropy of two probability densities, p and Π, measures the lack of information in Π relative to p and is given by

$$P(p, \Pi) = \int_{\mathbb{R}^N} p \ln\left(\frac{p}{\Pi}\right) d\vec{\lambda}. \tag{1.8}$$

In applications Π is called a prior distribution representing the prior state of knowledge; several applications will be developed below. The functional $P(p, \Pi)$ has several attractive properties,

(1.9)

- $P(p, \Pi) > 0$ unless $p \equiv \Pi$,
- $P(p, \Pi)$ is invariant under general changes of variables.

The key empirical idea of Jaynes is to use the measures of lack of information for probability densities in (1.7) and (1.8) to define the least biased probability measure for doing measurements on a physical system given by the information at hand. Thus, given some set of observed constraints defining a subset C of probability measures, the probability measure p^* for doing further measurements should be the one with the largest average lack of information, i.e. $p^* \in C$ satisfies the *maximum entropy principle*,

$$S(p^*) = \max_{p \in C} S(p). \tag{1.10}$$

Similarly, given some prior knowledge expressed through the probability measure Π for doing measurements, and the constraint set C from additional observations, the least biased probability distribution p^* for further measurements should minimize the lack of information in p given Π through (1.8), i.e. $p^* \in C$ satisfies the *minimum relative entropy principle*,

$$P(p^*, \Pi) = \min_{p \in C} P(p, \Pi). \tag{1.11}$$

Note that $P(p, \Pi)$ is a type of asymmetric distance function on pairs of probability measures also called the Kullback–Leibler distance. The formula in (1.8) readily generalizes to more general probability measures, dp and $d\Pi$, where dp is absolutely continuous with respect to $d\Pi$, i.e.

$$dp = \rho(\vec{\lambda})\, d\Pi, \quad \rho(\vec{\lambda}) \geq 0 \tag{1.12}$$

so that

$$P(p, \Pi) = \int \rho(\vec{\lambda}) \ln \rho(\vec{\lambda})\, d\Pi \tag{1.13}$$

with

$$\int \rho(\vec{\lambda})\, d\Pi = 1. \tag{1.14}$$

From these formulas it is evident that $P(p, \Pi)$ is an "inhomogeneous" version of $-S(p)$ taking into account the prior distribution $\Pi(\vec{\lambda})$.

Next we make a simple application of these ideas by utilizing the Gaussian density defined in (1.5) and (1.6). The central limit theorem vividly highlights the Gaussian distribution as the universal limiting distribution among probability densities with the same mean and second moment. Consider the constraint set C for probability measures $p(\lambda)$ defined by these first and second moments, i.e.

$$C = \left\{ p \in PM, \int \lambda p = \bar{\lambda}, \int (\lambda - \bar{\lambda})^2 p = \sigma^2 \right\}. \tag{1.15}$$

Given the universality of the Gaussian distribution, intuitively, the probability measure p^* satisfying the maximum entropy principle in (1.10) should be the Gaussian distribution, i.e. $p^* = p_{\bar{\lambda},\sigma}$. Next, we supply a simple direct proof by utilizing the relative entropy, $P(p, \Pi)$. For such a probability distribution p satisfying (1.15) with $p \neq p_{\bar{\lambda},\sigma}$, the first property in (1.9) guarantees that

$$\begin{aligned} 0 < P(p, p_{\bar{\lambda},\sigma}) &= \int p \ln p - \int p \ln p_{\bar{\lambda},\sigma} \\ &= \int p \ln p - \int p \ln(\sigma/\sqrt{2\pi}) - \int p \frac{(\lambda - \bar{\lambda})^2}{\sigma^2} \\ &= \int p \ln p - \int p_{\bar{\lambda},\sigma} \ln(\sigma/\sqrt{2\pi}) - \int p_{\bar{\lambda},\sigma} \frac{(\lambda - \bar{\lambda})^2}{\sigma^2} \\ &= S(p_{\bar{\lambda},\sigma}) - S(p), \end{aligned} \tag{1.16}$$

so that $S(p_{\bar{\lambda},\sigma}) > S(p)$ as required. The crucial third equality in (1.16) is true because, according to (1.15), p and $p_{\bar{\lambda},\sigma}$ have the same first and second moments.

A useful generalization of the simple fact just developed above involves finding the probability measure, $p^*(\vec{\lambda})$, with the least bias for doing further measurements given the constraints of several observed expected values,

$$\langle E_j \rangle = \int E_j(\vec{\lambda}) p(\vec{\lambda}) \, d\vec{\lambda}, \quad 1 \leq j \leq L, \tag{1.17}$$

so that

$$C_L = \{p \in PM, \langle E_j \rangle = \bar{E}_j, 1 \leq j \leq L\}. \tag{1.18}$$

A standard Lagrange multiplier calculation establishes that, formally, the least biased probability density $p^*(\vec{\lambda})$ for doing additional measurements given the ones defining C_L and satisfying the maximum entropy principle in (1.7) is given by

$$p_L^*(\vec{\lambda}) = C_{\vec{\theta}}^{-1} \exp\left(- \sum_{j=1}^{L} \theta_j E_j(\vec{\lambda}) \right) \tag{1.19}$$

provided that

$$C_{\vec{\theta}} = \int_{\mathbb{R}^N} \exp\left(- \sum_{j=1}^{L} \theta_j E_j(\vec{\lambda}) \right) d\vec{\lambda} < \infty. \tag{1.20}$$

In (1.19) and (1.20) θ is the vector of Lagrange multipliers guaranteeing $\langle E_j \rangle = \bar{E}_j$, $1 \leq j \leq L$. Many explicit examples of this standard Lagrange multiplier calculation are presented in [**31**]. We use this formalism in discussing equilibrium statistical mechanics for complex nonlinear dynamical systems below.

Next we use the above setting to give a simple demonstration of "coarse-graining". Intuitively, if $p^*_{L_1}(\vec{\lambda})$ is the probability density defined by the smaller number of constraints, $\langle E_j \rangle = \bar{E}_j$, $1 \le j \le L_1$ with $L_1 < L$, then we anticipate that this "coarse-grained" probability density, $p^*_{L_1}$, associated with fewer measurements, than p^*_L, should satisfy

$$P(p, p^*_{L_1}) > P(p, p^*_L) \quad \text{for any } p \in C, \tag{1.21}$$

so that the lack of information in $p^*_{L_1}$ exceeds that of p^*_L for any $p \in C$. Below we demonstrate the quantitative identity

$$P(p, p^*_{L_1}) = P(p, p^*_L) + P(p^*_L, p^*_{L_1}) \quad \text{for any } p \in C, \tag{1.22}$$

so that the loss of information through this coarse-graining procedure is given precisely by $P(p^*_L, p^*_{L_1})$. As in the Gaussian example above, we use the explicit form for p^*_L in (1.19) and calculate for any number of constraints $\tilde{L}$ in (1.18) and any p satisfying these constraints,

$$\begin{aligned} P(p, p^*_{\tilde{L}}) &= -S(p) - \int p \ln p^*_{\tilde{L}} \\ &= -S(p) + \ln C_{\vec{\theta}} \int p + \sum_{j=1}^{\tilde{L}} \theta_j \int E_j p \\ &= -S(p) + \ln C_{\vec{\theta}} \int p^*_{\tilde{L}} + \sum_{j=1}^{\tilde{L}} \theta_j \int E_j p^*_{\tilde{L}} \\ &= -S(p) + S(p^*_{\tilde{L}}). \end{aligned} \tag{1.23}$$

With the identity in (1.23) for any $\tilde{L} \le L$ the result in (1.22) follows immediately.

The simple ideas illustrated above have been utilized extensively by two of the authors to quantify the uncertainty in ensemble predictions for complex dynamical systems and, in particular, the non-Gaussian features in such ensemble predictions. Section 1.6 below provides a brief introduction to this important topic with many additional references listed there for the interested reader.

1.2. Equations for basic geophysical flows

The simplest geophysical model introduced here involves barotropic flow in periodic geometry with topography, forcing and dissipation. The equations are given by

$$\begin{aligned} q &= \Delta\psi + h, \\ \vec{v} &= \nabla^\perp \psi = \begin{pmatrix} -\psi_y \\ \psi_x \end{pmatrix}, \\ \frac{\partial q}{\partial t} &+ \vec{v} \cdot \nabla q = D(\Delta)\psi + F. \end{aligned} \tag{1.24}$$

In (1.24), q is the potential vorticity, $\omega = \Delta\psi$ is the relative vorticity, $\vec{v}$ is the incompressible fluid velocity, and h is the prescribed topography, with $D(\Delta)\psi$ being the dissipation and F the external forcing. The simplest example of a dissipation operator for such geophysical flows is Newtonian (eddy) viscosity in the form

$$D(\Delta)\psi = d\Delta^2\psi, \quad d > 0, \tag{1.25}$$

although many other physical dissipation mechanisms such as Ekmann friction arise in practice. When the topography, h, vanishes, the equations in (1.24) with (1.25) become the Navier-Stokes equations for a two-dimensional incompressible fluid in the vorticity-stream form [**26**]. However, topography introduces one important geophysical effect which quantitatively changes the nature of solutions of (1.24). More detailed discussion for these topics as well as the material presented below can be found in [**31**, Chapter 1].

1.2.1. Conserved quantities without forcing and dissipation. If we consider the equations in (1.24) without dissipation and forcing so that $D(\Delta) \equiv 0$ and $F \equiv 0$, the resulting equations are given by

$$\begin{aligned} q &= \Delta\psi + h, \\ \frac{\partial q}{\partial t} &+ \nabla^{\perp}\psi \cdot \nabla q = 0. \end{aligned} \tag{1.26}$$

The equations in (1.26) have an infinite number of conserved quantities for smooth solutions. First, the kinetic energy,

$$E = \frac{1}{2}\int |\vec{v}|^2 = -\frac{1}{2}\int \psi(q-h), \tag{1.27}$$

is conserved in time. Also the so-called generalized enstrophies,

$$\int G(q), \tag{1.28}$$

are conserved in time for any smooth $G(q)$. In particular, the integral of any power of potential vorticity is conserved,

$$Q_p(q) = \int q^p, \quad p = 1, 2, 3, \ldots. \tag{1.29}$$

In the above, $\int$ denotes the integral over the periodic domain where automatically $Q_1 \equiv 0$. The quadratic conserved quantity is also called the enstrophy $\mathcal{E}$ with

$$\mathcal{E} = \tfrac{1}{2}Q_2(q). \tag{1.30}$$

Clearly, geophysical flows are a rich dynamical system with infinitely many conserved quantities. The statistical significance of these conserved quantities is of great importance and is discussed next.

1.3. Information theory and equilibrium statistical mechanics for complex nonlinear systems

First we consider a general system of ODEs

$$\frac{d\vec{X}}{dt} = \vec{F}(\vec{X}) \tag{1.31}$$

with $\vec{X} \in \mathbb{R}^N$ and sketch the fashion in which the ideas from Section 1.1 can be utilized to predict long time measurements for (1.31). The key additional property needed for the system in (1.31) in order to do this is the Liouville property,

$$\operatorname{div}_{\vec{X}} \vec{F} = 0, \tag{1.32}$$

i.e. the vector field $\vec{F}$ is incompressible in phase space. One of the themes of the present book is the need to consider statistical solutions of (1.31) when the dynamics is chaotic and strongly nonlinear. Thus, we consider probability densities

$p(\vec{X})$ on $\mathbb{R}^N$ rather than individual solutions. An initial probability density $p_0(\vec{X})$ evolves in time to a probability density $p_t(\vec{X})$ which satisfies the Liouville equation,

$$\begin{aligned}\frac{\partial}{\partial t}p_t(\vec{X}) + \vec{F}\cdot\nabla_{\vec{X}}p_t(\vec{X}) = 0,\\ p_t\big|_{t=0} = p_0(\vec{X}).\end{aligned} \tag{1.33}$$

With the Liouville property in (1.32) and the transport equation in (1.33) for statistical solutions, there are an infinite number of conserved quantities for sufficiently smooth decaying solutions given by

$$\int G(p_t(\vec{X}))\,d\vec{X} = \int G(p_0(\vec{X}))\,d\vec{X} \tag{1.34}$$

for all times and any nice function $G(p)$.

As illustrated earlier in Section 1.2 for simple geophysical flows, physical systems often have additional conserved quantities, $E_j(\vec{X})$, $1 \le j \le L$, which satisfy

$$\frac{d}{dt}E_j\big(\vec{X}(t)\big) = 0, \quad 1 \le j \le L \tag{1.35}$$

for any solution $\vec{X}(t)$ of (1.31). Associated with such a family of conserved quantities in the deterministic system is the family of expected values for statistical solutions

$$\langle E_j\rangle_p = \int E_j(\vec{X})p(\vec{X})\,d\vec{X}, \tag{1.36}$$

which satisfy the conservation principle,

$$\langle E_j\rangle_{p_0} = \langle E_j\rangle_{p_t} \tag{1.37}$$

for all times t.

Now, given the nonlinear dynamical system in (1.31) satisfying the Liouville property, the question arises for how to determine a probability measure for doing long term statistical measurements. We apply the ideas from information theory developed in Section 1.1. Thus, consider a constraint set C for probability densities associated with the conserved quantities E_j, $1 \le j \le L$ so that

$$C = t\{p(\vec{X}) \in PM, \langle E_j\rangle_p = \bar{E}_j, 1 \le j \le L\}. \tag{1.38}$$

First, from the property in (1.37) the set C is invariant in time under the dynamics of statistical solutions. Secondly, by utilizing the special nonlinear function, $G(p) = -p\ln p$ in the conservation principle in (1.34), the entropy $S(p)$ from (1.7) is also conserved in time,

$$S(p_t) = S(p_0) \tag{1.39}$$

for all times t. Next, we choose $p^*(\vec{X})$ as the least biased probability measure for doing measurements given the constraints C in (1.38). With (1.19) and (1.20) this probability density is given by

$$p^*(\vec{X}) = \mathcal{G}_{\vec{\theta}}(\vec{X}) = C_{\vec{\theta}}^{-1}\exp\left(-\sum_{j=1}^{L}\theta_j E_j(\vec{X})\right) \tag{1.40}$$

provided that

$$C_{\vec{\theta}} = \int_{\mathbb{R}^N} \exp\left(-\sum_{j=1}^{L} \theta_j E_j(\vec{X})\right) d\vec{X} < \infty. \tag{1.41}$$

In (1.40) we have denoted $p^*(\vec{X})$ by $\mathcal{G}_{\vec{\theta}}(\vec{X})$ to emphasize that $p^*(\vec{X})$ becomes the canonical Gibbs measure in the setting of classical statistical mechanics [**43**]. This measure $p^*(\vec{X})$ is an invariant measure for the statistical dynamics for (1.31) and satisfies

$$\vec{F} \cdot \nabla_{\vec{X}} p^* = 0.$$

The complete details of the above discussion can be found in [**31**, Chapter 6]. By combining all of the above facts together with the calculation in (1.23), it is easy for the reader to check that

$$P(p_0, \mathcal{G}_{\vec{\theta}}) = P(p_t, \mathcal{G}_{\vec{\theta}}) = S(\mathcal{G}_{\vec{\theta}}) - S(p_t) = S(\mathcal{G}_{\vec{\theta}}) - S(p_0) \tag{1.42}$$

for all times t and any $p_0 \in C$. Furthermore, as in Section 1.1, we can quantify the coarse-grained information loss in utilizing only a subset of the conserved quantities.

Given $\mathcal{G}_{\vec{\theta}}(\vec{X})$ as the canonical measure for doing the least biased statistical measurements, we can make statistical predictions for the system in (1.31) by invoking an *ergodic hypothesis*, i.e.

For typical trajectories $\vec{X}(t)$ and a general functional, $F(\vec{X})$,

$$\lim_{T\to\infty} \frac{1}{T} \int_{T_0}^{T_0+T} F(\vec{X}(t))\, dt = \langle F \rangle_{\mathcal{G}_{\vec{\theta}}} = \int F(\vec{X}) \mathcal{G}_{\vec{\theta}} d\vec{X}. \tag{1.43}$$

The important point in (1.43) is that the left hand side can be evaluated experimentally by calculating the statistics of a (numerically) computed typical trajectory of the nonlinear dynamical system in (1.31) and the statistics for the Gibbs measure, $\mathcal{G}_{\vec{\theta}}(\vec{X})$ on the right hand side leads to practical predictions of long term (climate) behavior a priori! Nontrivial examples are presented below in Section 2.6 of Chapter 2 with much more found in [**31**] and the references listed there. The reader might already be wondering about the relationship between the dynamics for systems like (1.31) with the Liouville property and systems with dissipation and forcing: this issue is discussed briefly in Section 1.4 below and for an instructive highly nontrivial example in Section 2.8 below. Next, we briefly discuss a series of examples.

1.3.1. A simple example violating the Liouville property. Clearly, the Liouville property in (1.32) for the vector field $\vec{F}$ in (1.31) is central for the above theory. To see the importance of this assumption, it is interesting to see if there are dynamical systems which have a conserved quantity, but violate the Liouville property in (1.32) and have no nontrivial invariant measures which are absolutely continuous with respect to the Lebesgue measure. The following dynamical system in the plane provides such an example. Let $\vec{X} = (X_1, X_2)^T$ and $F^T = X_1 \vec{X}^\perp = (-X_1 X_2, X_1^2)$, i.e. consider the dynamical system

$$\frac{dX_1}{dt} = -X_1 X_2, \quad \frac{dX_2}{dt} = X_1^2. \tag{1.44}$$

This system has the family of conserved quantities

$$E(\vec{X} \cdot \vec{X}) \tag{1.45}$$

for an arbitrary smooth function, $E(s)$, since it is elementary to check that the trajectories for (1.44) lie on circles enclosing the origin. Also the vector field $\vec{F}$ violates the Liouville property since

$$\operatorname{div}_{\vec{X}} \vec{F} = -X_2 \neq 0. \tag{1.46}$$

Statistical solutions of this equation satisfy the Fokker–Planck equation,

$$\frac{\partial p_t}{\partial t} + \operatorname{div}_{\vec{X}}(\vec{F} p_t) = 0, \tag{1.47}$$

and invariant measures like the Gibbs measure which are a function of the conserved quantities necessarily are steady solutions of (1.47), i.e.

$$\operatorname{div}_{\vec{X}}(\vec{F} p) = 0. \tag{1.48}$$

Now, it is an exercise for the reader to check that for any smooth conserved quantity $E(\vec{X} \cdot \vec{X})$ from (1.45),

$$\operatorname{div}_{\vec{X}}(\vec{F} E) \neq 0, \tag{1.49}$$

so that no nontrivial function of the conserved quantities defines an invariant measure; in fact, the reader easily checks that the invariant measures satisfying (1.48) in the sense of distributions are not absolutely continuous with respect to Lebesgue measure and given by

$$\delta(X_1) \otimes \rho(X_2), \quad \rho \in PM(\mathbb{R}). \tag{1.50}$$

1.3.2. Statistical dynamics for truncated geophysical flows. It is not possible to apply statistical mechanics directly in infinite dimensions to the geophysical flow equations in (1.26) without damping and forcing. Thus, it is very natural to discuss these issues for judiciously chosen finite-dimensional approximations to the equations in (1.26).

The simplest such approximation is the

1.3.2.1. *Traditional spectral truncation.* We truncate the equations in (1.26) by projecting them onto the finite $(2M+1) \times (2M+1)$ set of Fourier modes. With help of the projection operator

$$P_\varepsilon f(\vec{x}) = \sum_{\substack{|k_1| \\ |k_2| \leq M}} \hat{f}_{\vec{k}} e^{i\vec{x}\cdot\vec{k}}, \quad \varepsilon = \frac{2\pi}{2M+1}, \tag{1.51}$$

where we adopt the notation

$$P_\varepsilon f \equiv f^\varepsilon,$$

we write the standard Galerkin projection of the quasi-geostrophic equations in (1.26) as

$$\begin{aligned} \frac{\partial q^\varepsilon}{\partial t} &= -P_\varepsilon[\operatorname{div}(\vec{v}^\varepsilon q^\varepsilon)], \\ q^\varepsilon &= \Delta\psi^\varepsilon + h^\varepsilon, \\ \vec{v}^\varepsilon &= \nabla^\perp \psi^\varepsilon. \end{aligned} \tag{1.52}$$

The following three integrals of motion survive the truncation: the projected spatial integral of the potential vorticity Q_1^ε, the projected energy E^ε, and the projected enstrophy $\mathcal{E}^\varepsilon$. In the framework of (1.51) and (1.52) the proofs of conservation are trivial: taking into account the fact that for the standard L^2-inner product $(g^\varepsilon, f) = (g, f^\varepsilon)$ for all g, f, we write

- Vorticity

$$\frac{\partial Q_1^\varepsilon}{\partial t} = \int_{T^2} \frac{\partial q^\varepsilon}{\partial t} = -\int_{T^2} \operatorname{div} P_\varepsilon(\vec{v}^\varepsilon q^\varepsilon) = 0, \tag{1.53}$$

- Energy

$$\begin{aligned} \frac{\partial E^\varepsilon}{\partial t} &= -\int_{T^2} \psi^\varepsilon \frac{\partial q^\varepsilon}{\partial t} = \int_{T^2} \psi^\varepsilon P_\varepsilon[\operatorname{div}(\vec{v}^\varepsilon q^\varepsilon)] \\ &= \int_{T^2} \psi^\varepsilon \operatorname{div}(\vec{v}^\varepsilon q^\varepsilon) = \int_{T^2} \operatorname{div}(\psi^\varepsilon \vec{v}^\varepsilon q^\varepsilon) = 0, \end{aligned} \tag{1.54}$$

- Enstrophy

$$\begin{aligned} \frac{\partial \mathcal{E}^\varepsilon}{\partial t} &= \int_{T^2} q^\varepsilon \frac{\partial q^\varepsilon}{\partial t} = -\int_{T^2} q^\varepsilon P_\varepsilon[\operatorname{div}(\vec{v}^\varepsilon q^\varepsilon)] \\ &= -\int_{T^2} q^\varepsilon \operatorname{div}(\vec{v}^\varepsilon q^\varepsilon) = -\frac{1}{2}\int_{T^2} \operatorname{div}(q^\varepsilon \vec{v}^\varepsilon q^\varepsilon) = 0. \end{aligned} \tag{1.55}$$

We note that the Galerkin truncation in (1.52) of the quasi-geostrophic equations in (1.26) can be interpreted as Fourier space equations limited to the finite $(2M+1)\times(2M+1)$ domain of the Fourier coefficients:

$$\frac{d\hat{q}_k}{dt} = \sum_{k_1', k_2' = -M}^{M} \frac{\vec{k} \times \vec{k}'}{|\vec{k}'|^2} \hat{q}_{\vec{k}+\vec{k}'} (\hat{q}_{-\vec{k}'} - \hat{h}_{-\vec{k}'}), \tag{1.56}$$

provided that we extend the finite domain of the Fourier coefficients up to $(4M+1)\times(4M+1)$ by using the periodicity rule

$$\hat{q}_{\vec{k}} = \hat{q}_{(\vec{k}+2M+1)}. \tag{1.57}$$

It is easy to check that this system satisfies the Liouville property so that equilibrium statistical mechanics can be applied to this truncated problem. The vorticity has $Q_1^\varepsilon = 0$ in periodic geometry so that only the two quadratic conserved quantities given by the energy, E, and enstrophy, $\mathcal{E}$, are important here. Applying the general theory developed in (1.31)–(1.43) of this section to statistical solutions, we obtain the family of Gibbs measures

$$\mathcal{G}_{\alpha,\mu} = C_{\alpha,\mu} \exp[-\alpha(\mathcal{E}^\varepsilon + \mu E^\varepsilon)] \tag{1.58}$$

provided the Lagrange multipliers, μ, α, satisfy $-1 < \mu < \infty$, $0 < \alpha$; these are Gaussian measures as given, for example, in (1.5) and (1.6). These Gaussian measures have a striking mean characterized by the solution, $\bar{\psi}_\mu$, of the linear equation

$$\Delta\bar{\psi}_\mu + h^\varepsilon = \mu\bar{\psi}_\mu. \tag{1.59}$$

Thus, remembering the ergodic hypothesis in (1.43), the statistical theory for these truncated equations predicts a striking nontrivial (climatological) mean state locked to the topography through (1.59). The original prediction is due to Salmon et al. [**39**] (also see [**38**]), and numerical experiments strikingly confirming this prediction are found in [**31**, Chapter 8] with many more details, references and further material.

Despite all of the interesting applications of this simplest statistical theory, it has some features that are fundamentally unsatisfactory. We have noted in Section 1.2 that the geophysical model in (1.26) has infinitely many conserved quantities beyond those of energy, E, and enstrophy, $\mathcal{E}$, but only these two survive the spectral truncation. A very interesting open problem is the role of such additional conserved quantities for the statistical dynamics of ensembles of solutions. This has been an outstanding open problem for nearly twenty-five years in statistical fluid dynamics; fundamental progress on these issues has been achieved very recently by two of the authors [**2, 3**]. This is described briefly next with further details developed in [**31**, Chapter 14].

1.3.2.2. *The spectral truncation with many conserved quantities.* Here, following the important observation by Zeitlin [**46**], we consider the sine-bracket truncation as an approximation to the geophysical dynamics in (1.26). This finite dimensional set of equations for the Fourier coefficients is given by

$$\frac{d}{dt}\hat{q}_{\vec{k}} = \sum_{k'_1,k'_2=-M}^{M} \frac{1}{\varepsilon}\frac{\sin(\varepsilon\vec{k}\times\vec{k}')}{|\vec{k}'|^2}\hat{q}_{\vec{k}+\vec{k}'}(\hat{q}_{-\vec{k}'} - \hat{h}_{-\vec{k}'}), \quad \varepsilon = \frac{2\pi}{2M+1}, \tag{1.60}$$

with the same convention of periodicity as in (1.57). This truncation conserves the energy in (1.54) as the Hamiltonian. In addition, the sine-bracket truncation in (1.60) conserves $2M$ independent Casimir invariants of the form

$$\begin{gathered} C_N = \sum_{Z^N} \hat{q}_{\vec{i}_1}\cdots\hat{q}_{\vec{i}_N}\cos[\varepsilon A(\vec{i}_1,\ldots,\vec{i}_N)], \\ Z^N = \left\{(\vec{i}_1,\ldots,\vec{i}_N), \sum_{j=1}^{N}\vec{i}_j = \vec{0}\right\}, \\ 1 \le N \le 2M; \\ A(\vec{i}_1,\ldots,\vec{i}_N) = \vec{i}_2\times\vec{i}_1 + \vec{i}_3\times(\vec{i}_1+\vec{i}_2) + \cdots + \vec{i}_N\times(\vec{i}_1+\cdots+\vec{i}_{N-1}). \end{gathered} \tag{1.61}$$

The Casimir invariant C_2 is a multiple of the enstrophy since $A(\vec{i},-\vec{i}) = 0$. However, the higher Casimir invariants C_N for $3 \le N \le 2M$ are suitable regularizations of the higher powers of potential vorticity from (1.29) for $3 \le N \le 2M$. We note that as the number of modes, M, increases,

$$\frac{\sin(\varepsilon\vec{k}\times\vec{k}')}{\varepsilon} = \vec{k}^{\perp}\cdot\vec{k}' + \mathcal{O}(\varepsilon^2),$$

so that by comparing (1.60) and the spectral representation, the sine-bracket truncation is formally consistent with the geophysical equations as $M\to\infty$. In fact, a rigorous proof of convergence is presented in [**2**] as well as further theoretical and computational discussion of (1.60), (1.61).

With this background, two of the authors [**3**] utilized an appropriate symplectic integrator with 512 total modes and definitely established the statistical significance of the regularized third power of potential vorticity,

$$C_3(q),$$

defined above. In particular, the mean state at large scales satisfied a nonlinear equation for $C_3(q)\neq 0$ rather than the simpler linear equation in (1.59); we discuss the significance of this and related results briefly in the Section 1.4 below.

To directly assess the statistical significance of all the conserved quantities in (1.61) is a formidable task in general. Clearly, it is desirable to have a simpler model where such issues can be discussed completely in a more transparent fashion. Next, we briefly discuss such a model. It will be utilized extensively in Section 2.6 of Chapter 2 in our discussion of the fluctuation-dissipation theorem.

1.3.3. The truncated Burgers–Hopf model (TBH). The inviscid Burgers–Hopf equation

$$\frac{\partial}{\partial t}u + \frac{1}{2}(u^2)_x = 0, \tag{1.62}$$

is one of the extensively studied models in applied mathematics. One prominent feature of this equation is the formation and propagation of shocks (discontinuity) from smooth initial data. Here we consider spectral truncations of the Burgers–Hopf equation (1.62). Such a numerical approximation would be disastrous if we are interested in the dynamics of (1.62) since shock waves form. However, our interest here, as developed in [**28, 29**], is models that capture certain salient features of atmosphere/ocean dynamics, and not that of (1.62). Suppose the Burgers–Hopf equation is equipped with periodic boundary conditions so that Fourier series may be applied. Let $P_\Lambda f = f_\Lambda$ denote the finite Fourier series truncation of f, i.e.,

$$P_\Lambda f = f_\Lambda = \sum_{|k|\leq\Lambda} \hat{f}_k e^{ikx}. \tag{1.63}$$

Here and elsewhere in the section it is tacitly assumed that f is 2π-periodic and real-valued so that the complex Fourier coefficients $\hat{f}_k$ satisfy $\hat{f}_{-k} = \hat{f}_k^*$, where $*$ denotes complex conjugation in this context. The positive integer, Λ, from (1.63) defines the number of complex-valued degrees of freedom in the approximation. With these preliminaries, the model introduced and studied here is the approximation to (1.62)

$$(u_\Lambda)_t + \tfrac{1}{2}P_\Lambda(u_\Lambda^2)_x = 0. \tag{1.64}$$

This is the Fourier Galerkin truncated approximation to (1.62).

It is elementary to show that solutions of the equations in either (1.64) conserve both momentum and energy, i.e.,

$$M = \int u_\Lambda(t) = \hat{u}_0 \tag{1.65}$$

and

$$E = \frac{1}{2}\int u_\Lambda^2 dx = \frac{1}{2}|\hat{u}_0|^2 + \sum_{k=1}^{\Lambda} |\hat{u}_k|^2 \tag{1.66}$$

are constant in time for solutions of (1.64). In addition to the conserved quantities in (1.65) and (1.66), it is easy to check that the third moment

$$H = \int u_\Lambda^3 dx \tag{1.67}$$

is also conserved. The quantity, H, is actually the Hamiltonian for the truncated Burgers–Hopf system [**1**].

It is well known that nontrivial smooth solutions of (1.62) develop discontinuities in finite time and thus exhibit a transfer of energy from large scales to small scales. However, once this transfer develops in a general solution of TBH, the conservation of energy constraint rapidly redistributes the energy in the smaller scales to the larger scale modes. This effect is responsible for the intuitive fact that the small scale modes of the system should decorrelate more rapidly than the large scale modes. This is a key feature of this system which satisfies the Liouville property. First, assume that the momentum, $M_0 = 0$, and that the energy E in (1.66) is the relevant statistical quantity for further measurements. Combining the Liouville property together with the conserved energy and the general framework for statistical solutions discussed earlier in this section, the canonical Gibbs measure is given by

$$\mathcal{G}_\theta = C_\theta^{-1} \exp\left(-\frac{\theta}{2} \sum_{k=1}^{\Lambda} |\hat{u}_k|^2 \right), \quad \theta > 0. \tag{1.68}$$

This is a product of identical Gaussian measures for each of the Fourier coefficients.

Next we ask whether the same procedure would work for the conserved quantity given by $H = \int u_\Lambda^3 dx$ from (1.67). This is an odd function of u_Λ and grows cubically at infinity so that the associated Gibbs measure is infinite. Thus, this conserved quantity provides an example which is not statistically normalizable even though it is the Hamiltonian for this system. We discuss the statistical relevance of this conserved quantity for the dynamics briefly below [**1**]. On the other hand, numerical simulations of TBH with suitable random initial data [**28, 29**] strongly confirm the equilibrium statistical predictions for TBH even though the effect of the Hamiltonian, H, is not taken into account in the Gibbs measure. Thus, as for the more complex geophysical flows, the question arises in this setting regarding the statistical significance of the conserved quantity H beyond the statistical significance of E. Clearly, this is a simplified context for the same issues which we discussed in detail earlier in this section for the more complex truncated geophysical equations. The advantages of such a simpler model is that detailed equilibrium statistical predictions beyond those utilized above involving both E and H can be made and compared directly to numerical simulations. This is done in [**1**]. Below is a brief summary of the main results and the reader is referred to that paper as well as in [**2**, Section 5] for further details.

First, an appropriate equilibrium statistical theory is developed which includes both the energy and Hamiltonian by restricting to fixed energy surfaces in a microcanonical ensemble [**43**]. Then a convergent Monte-Carlo algorithm is developed for computing equilibrium statistical distributions. The probability distribution of the Hamiltonian on a microcanonical energy surface is studied through a Monte-Carlo algorithm and leads to the concept of statistically relevant and irrelevant values for the Hamiltonian. Empirical numerical estimates and simple analysis are combined to demonstrate that the statistically relevant values of the Hamiltonian have vanishingly small measure as the number of degrees of freedom increases with fixed mean energy. The predictions of the theory for relevant and irrelevant values for the Hamiltonian are confirmed through systematic numerical simulations. For statistically relevant values of the Hamiltonian, these simulations show a surprising spectral tilt rather than equipartition of energy. This spectral tilt is predicted and confirmed independently by Monte-Carlo simulations based on equilibrium statistical mechanics together with a heuristic formula for the tilt. In particular, Monte-Carlo simulations of the equilibrium statistical distribution combined with a simple scaling theory [**1**] predict a formula for the tilt in the spectrum for the statistical energy distribution for TBH which reflects the additional conserved quantity, H. This formula is

$$\frac{1}{2}\langle |\hat{u}_k|^2 \rangle = \frac{E}{\Lambda}\left(1 + \frac{8}{5}\frac{H^2}{E^3}\frac{(\Lambda+1)/2 - k}{\Lambda} \right). \tag{1.69}$$

Notice that H^2/E^3 is a nondimensional number reflecting the importance of the conserved quantity, H, relative to E and gives the slope of the spectral tilt beyond equipartition of energy; when this number is very small, essentially equipartition of energy is predicted. Many numerical experiments confirming the striking prediction in (1.69) are developed in [**1, 2**]. In particular, Abramov and Majda [**2**] show that the breakdown of (1.69) roughly coincides with the failure of mixing, i.e. non-decay of correlations at large scales, for very large values of H. A list of accessible open mathematical problems regarding statistically relevant conserved quantities for TBH is presented at the end of [**1**].

Finally, we end this section by indicating how we can use information theory to do equilibrium statistical mechanics for both the energy, E, and Hamiltonian, H. The idea is simple: since the energy is conserved, we can validly restrict the dynamics to a constant energy surface which in this case is a compact $(2\Lambda - 1)$-dimensional sphere, $S_{\bar{E},\Lambda}$, in the Fourier series representation. Let $d\nu_{\bar{E},\Lambda}$ denote this uniform probability measure on this sphere with radius $\sqrt{2\bar{E}}$. On this compact surface, $H(u_\Lambda)$ is finite and uniformly bounded, and therefore renormalizable. Then, as in (1.12) and (1.13) above, we seek a probability measure which is absolutely continuous with respect to the surface measure $d\nu_{\bar{E},\Lambda}$ and satisfies the minimum relative entropy principle in (1.11), and therefore has the least bias for doing further measurements. With (1.11), (1.12) and (1.13), the result is the mixed canonical-microcanonical Gibbs measure given by

$$\mathcal{G}_{\bar{E},\Lambda,\theta} = \frac{\exp(-\theta H(u_\Lambda))\, d\nu_{\bar{E},\Lambda}}{\int_{S_{\bar{E},\Lambda}} \exp(-\theta H(u_\Lambda))\, d\nu_{\bar{E},\Lambda}}. \tag{1.70}$$

The probability measure in (1.70) provides the theoretical framework for developing equilibrium Monte-Carlo simulations which, for example, predict the scaling law for energy in (1.69). See the paper of Abramov et al. [**1**] for the Monte-Carlo algorithm and the above results.

Note that a similar mixed microcanonical-canonical Gibbs ensemble can be developed for truncated geophysical flows with many conserved quantities. In these systems it is very natural to impose constant energy on a fixed microcanonical surface since the energy E for geophysical flows is a property of the large scales while conserved nonlinear functionals of q as in (1.61) involve smaller scale features of the flow and should be imposed by least-biased probability measures utilizing information theory as illustrated above for TBH.

1.4. A systematic mathematical strategy for coarse-grained equilibrium statistical predictions

In the previous section we illustrated how to build a complete equilibrium statistical theory for truncated geophysical flows. The result is an elegant statistical theory which can be used for predictions of the mean, variance and fluctuations at large scales. In many practical applications for geophysical flows one is primarily interested in the predicted mean state itself which defines the "climatology". For example, as illustrated in Section 1.3 above, the Gaussian energy-enstrophy theory for truncated geophysical flows yields the simple mean state prediction,

$$\Delta\bar{\psi}_\mu + h = \mu\bar{\psi}_\mu. \tag{1.71}$$

One way to simplify the theory and to make empirical statistical predictions for the mean state is to "coarse-grain" the statistics being measured to those appropriate only for individual pointwise values and ignore the spatial correlations. To develop these ideas, we consider the basic geophysical flows in (1.26) with topography and no dissipation and forcing in periodic geometry normalized to area 1; the conserved quantity given by the energy

$$E = \frac{1}{2}\int |\vec{v}|^2 = -\frac{1}{2}\int \psi(q-h) \tag{1.72}$$

will be important for the theory developed below since it involves the large scales and strongly influences the mean state. Recall that the potential vorticity is given by $q = \Delta\psi + h$.

With the above remarks, we consider a special family of probability measures on $T^2 \times \mathbb{R}$ with density $\rho(\vec{x}, \lambda) \geq 0$ so that

- For almost all $\vec{x} \in T^2$, $\rho(\vec{x}, \lambda)$ is a probability density on $\mathbb{R}$, i.e.
$$\int_{\mathbb{R}} \rho(\vec{x}, \lambda)\, d\lambda = 1.$$

(1.73)
- For almost all $\vec{x} \in T^2$ $\rho(\vec{x}, \lambda)$ is a probability density associated with the potential vorticity $q(\vec{x})$ at location $\vec{x}$, i.e.
$$\int_{q_-}^{q_+} \rho(\vec{x}, \lambda)\, d\lambda = \text{Prob}\{q_- \leq q(\vec{x}) \leq q_+\}.$$

By a slight abuse of notation we denote the set of such probability measures satisfying (1.73) as $PM(T^2 \times \mathbb{R})$. In mathematical terms, these are Young measures [**26, 42**]. Note that such a family of probability measures "coarse-grains" the general statistics of a random field for $q(\vec{x})$ on T^2 to only the uncorrelated single point probability densities. As discussed earlier at the beginning of Section 1.1, given any function $F(\vec{x}, q)$ of the potential vorticity, for the class of probability measures in (1.73) we calculate the expected value pointwise through the density $\rho(\vec{x}, \lambda)$ by

$$\bar{F}_\rho(\vec{x}) = \langle F(\vec{x}, q)\rangle_\rho = \int_{\mathbb{R}} F(\vec{x}, \lambda)\rho(\vec{x}, \lambda)\, d\lambda, \tag{1.74}$$

and the overall expected value by

$$\langle F\rangle_\rho = \int_{T^2} \langle F(\vec{x}, q)\rangle\, d\vec{x}. \tag{1.75}$$

In (1.75) remember that the period domain has area 1.

For the family of probability densities defined above, it is completely natural to extend the one-dimensional relative entropy in (1.8) to the special probability densities $\rho(\vec{x}, \lambda)$ by integration in $\vec{x}$. In other words, for $\rho(\vec{x}, \lambda)$, $\Pi_0(\vec{x}, \lambda) \in PM$, the functional

$$P(\rho, \Pi_0) = \int_{T^2} d\vec{x} \int_{\mathbb{R}} \rho(\vec{x}, \lambda) \ln\left(\frac{\rho(\vec{x}, \lambda)}{\Pi_0(\vec{x}, \lambda)}\right) d\lambda \tag{1.76}$$

measures the lack of information in the prior distribution Π_0 relative to ρ. Next we define the mean kinetic energy for a probability density $\rho(\vec{x}, \lambda)$ satisfying (1.73). Remembering (1.74), we calculate the mean potential vorticity by

$$\bar{q}(\vec{x}) = \int_{\mathbb{R}} \lambda\rho(\vec{x}, \lambda)\, d\lambda \tag{1.77}$$

and the associated mean stream function by

$$\Delta\bar{\psi} + h = \bar{q}(\vec{x}). \tag{1.78}$$

With these two formulas and (1.72), the mean kinetic energy for a probability density $\rho(\vec{x}, \lambda)$ is given by

$$E(\rho) = -\frac{1}{2}\int \bar{\psi}(\bar{q} - h). \tag{1.79}$$

Now we have all the ingredients to develop an empirical equilibrium statistical theory for making coarse-grained equilibrium predictions of the one-point statistics of geophysical flows at large scales. First, start with a judicious prior distribution $\Pi_0(\vec{x}, \lambda)$ which represents statistical fluctuations of potential vorticity at small scales. As discussed earlier, the energy, $\bar{E}$, reflects the large scale properties of the flow, so given the prior distribution, $\Pi_0(\vec{x}, \lambda)$, find the probability density $\rho^*(\vec{x}, \lambda) \in PM$ with the least bias for further measurements. In other words, define the constraint set C by

$$C = \{\rho \in PM(T^2 \times \mathbb{R}), E(\rho) = \bar{E}\}, \tag{1.80}$$

and choose the measure $\rho^*(\vec{x}, \lambda)$ belonging to C and satisfying

$$P(\rho^*, \Pi_0) = \min_{\rho \in C} P(\rho, \Pi_0). \tag{1.81}$$

With ρ^* we can make coarse-grained large scale predictions for the mean state through the two formulas,

$$\begin{aligned} q^*(\vec{x}) &= \int \lambda \rho^*(\vec{x}, \lambda)\, d\lambda, \\ \Delta\psi^* + h = q^* &= \int \lambda \rho^*(\vec{x}, \lambda)\, d\lambda. \end{aligned} \tag{1.82}$$

As developed in detail in [**31**, Chapter 9], the formulas in (1.82) can be written elegantly in terms of the partition function, $Z(\psi, \vec{x})$, also called the moment generating function of the prior probability distribution, $\Pi_0(\vec{x}, \lambda)$. The partition function $Z(\psi, \vec{x})$ is defined by

$$Z(\psi, \vec{x}) = \int e^{\psi\lambda}\Pi_0(\vec{x}, \lambda)\, d\lambda, \tag{1.83}$$

and (1.82) becomes the mean field equation

$$\Delta\psi^* + h = \frac{1}{\theta}\left(\frac{\partial}{\partial\psi} \ln Z(\theta\psi, \vec{x})\right)\bigg|_{\psi=\psi^*}. \tag{1.84}$$

In (1.84), θ is the Lagrange multiplier determined by the energy constraint, $E(\rho) = \bar{E}$; thus, the mean field equation is a nonlinear elliptic equation for ψ^* in terms of itself.

Clearly, the above theory is an empirical statistical theory for making large scale predictions of the mean state. The following questions arise immediately:

- Are interesting physical phenomena predicted by the theory?
- In a given physical problem, how do we pick the prior distribution, Π_0?
- How much of the theory applies to more realistic problems with dissipation and forcing in suitable regimes?
- Are there a priori mathematical and computational confirmation of this coarse-grained statistical strategy in suitable idealized circumstances?

All of these issues are discussed in detail in [**31**, Chapters 9–14], where extensive references are given to additional literature. Here we only briefly list some highlights and provide some intuition in our discussion. First, there are many competing complete equilibrium statistical theories such as the energy-enstrophy theory described in Section 1.3, the point vortex theory, and others. It is shown in [**31**, Chapter 9] that the predictions of each of these theories can be recovered by a suitable choice of the prior distribution, $\Pi_0(\lambda)$, in the coarse-grained simplified empirical approach developed here. To illustrate this, we claim that the simplest Gaussian prior distribution for small scale fluctuations of potential vorticity,

$$\Pi_0(\lambda) = \sqrt{\frac{\alpha}{2\pi}} e^{-\alpha\lambda^2/2}, \tag{1.85}$$

recovers the prediction of the energy-enstrophy statistical theory in (1.71). With (1.85) it is easy to calculate that

$$Z(\psi) = e^{\psi^2/2\alpha}, \tag{1.86}$$

so that (1.84) becomes

$$\Delta\psi^* + h = \frac{\theta}{\alpha}\psi^*, \tag{1.87}$$

and identifying $\mu = \theta/\alpha$ we recover (1.71).

Recall that the Gaussian distribution is the simplest and least biased prior distribution using the first and second moments alone. In practice, prior distributions for small scale fluctuations of potential vorticity should also be skewed and even with long tails on one side to reflect a preponderance of either cyclones or anticyclones in small scale potential vorticity fluctuations in a given physical context. Thus, it is interesting to build a family of prior probability distributions $\Pi_0(\lambda)$ measuring deviations from a Gaussian distributions through skewness, the third moment of $\Pi_0(\lambda)$, and also with a higher probability of large amplitude events going in the same direction as this skewness. With the normalizations for the first and second moments,

$$\int \lambda\Pi_0(\lambda)\, d\lambda = 0, \qquad \int \lambda^2\Pi_0(\lambda)\, d\lambda = 1,$$

the skewness is given by the third moment

$$\int \lambda^3\Pi_0(\lambda)\, d\lambda = \operatorname{Skew}\Pi_0. \tag{1.88}$$

An attractive choice of a family of prior distributions fitting these requirements is the family of centered Γ-distributions,

$$\Pi_0^\varepsilon(\lambda) = |\varepsilon|^{-1} R(\varepsilon^{-1}[\lambda + \varepsilon^{-1}]; \varepsilon^{-2}), \tag{1.89}$$

where $R(z;a) = \Gamma(a)^{-1} z^{a-1} e^{-z}$ $(z \geq 0)$, $R(z;a) = 0$ $(z < 0)$ denotes the standard Γ-density. This family has a one-sided exponential tail and compact support on the other side, with the Γ-distribution shifted to have zero mean. By construction, $\Pi_0^\varepsilon(\lambda)$ satisfies the normalizations in (1.88) with

$$\operatorname{Skew}\Pi_0^\varepsilon = 2\varepsilon. \tag{1.90}$$

Since the Γ-distribution, $R(z;a)$, has the explicit generating or partition function, $Z(\psi, a) = (1-\psi)^{-a}$, it is not difficult to show that in this case, the mean field equation in (1.84) is given by

$$\Delta\psi^* + h = \frac{\theta\psi^*}{1 - \varepsilon\theta\psi^*}, \tag{1.91}$$

so that we recover the Gaussian prediction for $\varepsilon = 0$.

The numerical simulations of Abramov and Majda [**3**] for the truncated geophysical flow with many conserved quantities provide a striking confirmation of the trends predicted by these empirical statistical theories discussed here. In the simulations with a nontrivial regularized third power of the potential vorticity, as this conserved quantity is varied, it is established that the small scale one-point statistics for potential vorticity have a probability distribution well-approximated by the skewed prior distribution Π_0^ε, while the coarse-grained large scale mean is simultaneously well-approximated by solutions of (1.91) with same skewness parameter ε! (See [**31**, Chapter 14] for more details.) Such an approach is also utilized to successfully predict the Great Red Spot of Jupiter through a physically motivated prior distribution $\Pi_0^\varepsilon(\lambda)$ based on observations from the Galileo mission [**45**]. This is discussed briefly in the next section and also extensively in [**31**]. In the mathematical setting of appropriate idealized equilibrium statistical distributions with stochastic rearrangement dynamics rather than approximations for the actual fluid equations, there is a rigorous mathematical confirmation of the empirical statistical theory developed in this section provided there is a large separation of scales between the small scale potential vorticity fluctuations and the large scales containing energy [**7**].

Finally, the above statistical theory has been shown to apply to flows with random small scale forcing and dissipation provided the flux of energy to large scales is sufficiently weak through "crude closure" algorithms based on predicting changes of the empirical equilibrium statistical state with time [**11, 14, 15**]. There is even a recent rigorous mathematical proof of this behavior in an idealized setting [**30**]. All of this is discussed in detail in [**31**, Chapter 12].

1.5. Application to predicting Jupiter's Great Red Spot

Prominent examples of long lived large scale vortices in geophysical flows are those such as the Great Red Spot (GRS) on Jupiter [**12, 32, 36**]. The emergence and persistence of such coherent structures at specific latitudes such as $22.4°S$ for the GRS in a background zonal shear flow that seems to violate all of the standard stability criteria [**12**] are a genuine puzzle. In [**45**] the equilibrium statistical theory with a suitable prior distribution, introduced in Section 1.4, is utilized to predict the actual coherent structures on Jupiter in a fashion which is consistent with the known observational record. As discussed in Section 1.4, the statistical theory is based on a few judiciously chosen conserved quantities for the inviscid dynamics such as energy and circulation and does not involve any detailed resolution of the fluid dynamics. The key ingredient is the prior distribution, $\Pi(\lambda)$, for the one-point statistics of the potential vorticity which parameterizes the unresolved small scale turbulent eddies that produce the large scale coherent structures. It is shown in [**45**] how the observational record of Jupiter from the recent Galileo mission suggests a special structure for such a prior distribution resulting from intense small scale forcing. Recall, as discussed at the end of Section 1.4, that the equilibrium statistical theory can describe coherent structures occurring from strong small scale forcing provided that the flux of energy to large scales is sufficiently weak

[**11, 14, 15**]. With the confidence gained from this fact, a physically motivated skewed Γ prior distribution is utilized to predict the large scale coherent structures on Jupiter from the empirical equilibrium statistical theory discussed in Section 1.4; these predicted structures are in agreement with the observational record from the weather layer of Jupiter as inferred from the Voyager mission. The approach in making the theoretical prediction is inherently multi-disciplinary and has several ingredients:

A) The observational record for Jupiter.
B) The quasi-geostrophic model for interpreting observations and predictions.
C) Fitting the zonal shear profile from observations, the Limaye profile from the Voyager mission [**36**], to the quasi-geostrophic model [**12**].
D) The reinterpretation of C) using the energy-enstrophy statistical theory.
E) The observational and physical motivation for the prior distribution for the equilibrium statistical theory and a convenient mathematical framework for interpreting the statistical theory with this family of prior distributions.
F) Equilibrium statistical predictions for the coherent structures on Jupiter and comparison with observations.

This approach is taken in [**45**] and leads to a successful prediction of the location and structure of the Great Red Spot. It is discussed in [**31**, Chapter 13].

1.6. Information theory and predictability for complex nonlinear systems

Complex systems with many spatial degrees of freedom arise in environmental science in diverse contexts such as atmosphere/ocean general circulation models (GCMs) for climate or weather prediction [**18**], pollution models, and models for the spread of hazardous biological, chemical, or nuclear plumes. These nonlinear models are intrinsically chaotic over many time scales with sensitive dependence on initial conditions. In this chapter, such models are represented discretely as a large system of ODEs for a vector $\vec{X} \in \mathbb{R}^N$ given by

$$\frac{d\vec{X}}{dt} = \vec{F}(\vec{X}, t), \tag{1.92}$$

with $N \gg 1$.

Given both the uncertainty in a deterministic initial condition, $\vec{X}_0$, as well as the intrinsic chaos in solutions of (1.92), it is natural instead to consider an ensemble of initial data characterized by a probability density, $p_0(\vec{X})$, satisfying $p_0 \geq 0$, $\int p_0 = 1$ and with mean given by $\vec{X}_0$, i.e.

$$\int_{\mathbb{R}^N} \vec{X} p_0(\vec{X})\, d\vec{X} = \vec{X}_0. \tag{1.93}$$

The idea is to utilize the ensemble of solutions of (1.92) drawn from the initial data $p_0(\vec{X})$ to quantify the uncertainty and measure the confidence interval and predictive power of the deterministic solution beginning at $\vec{X}_0$. For example, consider $\vec{X}_0$, and an associated probability density consisting of small random perturbations of this initial data. This probability density measures the uncertainty in the measurement of the initial data. For example, the random initial data might be sampled from a Gaussian probability distribution centered about $\vec{X}_0$,

$$p_0(\vec{X}) = (2\pi)^{-N/2} \varepsilon^{-N/2} e^{-|\vec{X}-\vec{X}_0|^2/2\varepsilon}, \quad \varepsilon \ll 1 \tag{1.94}$$

where $\varepsilon \ll 1$ measures the variance in each component. In other words, the ensemble of initial data satisfies

$$\int \vec{X} p_0(\vec{X})\, d\vec{X} = \vec{X}_0,$$
$$\int (X_i - X_{0,i})(X_j - X_{j,0}) p_0(\vec{X})\, d\vec{X} = \varepsilon \delta_{ij} \tag{1.95}$$

where δ_{ij} are the Kronecker symbols and take the value 1 if $i = j$ and zero otherwise. In general, here we assume an initial probability measure $p_0(\vec{X})$, reflecting an ensemble of initial conditions.

Theoretically, this is straightforward because it is easy to establish that the initial probability density, $p_0(\vec{X})$, evolves to a new probability density, $p_t(\vec{X})$, which at later times satisfies the Liouville equation,

$$\frac{\partial}{\partial t} p_t(\vec{X}) + \operatorname{div}_{\vec{X}}(\vec{F} p_t) = 0,$$
$$p_t(\vec{X})\big|_{t=0} = p_0(\vec{X}). \tag{1.96}$$

The Liouville equation is a linear PDE in a very large spatial dimension, $\mathbb{R}^N$, with $N \gg 1$, and is impractical to solve directly. In practice, instead a finite ensemble of individual solutions of (1.92), $\{\vec{X}_t^r, 1 \le r \le R\}$, is constructed by sampling the initial distribution, $p_0(\vec{X})$. Then $p_t(\vec{X})$ is approximated by the empirical probability distribution for ensemble prediction

$$p_t^E = \frac{1}{R} \sum_{r=1}^{R} \delta_{\vec{X}_t^r}(\vec{X}) \tag{1.97}$$

where $\delta_{\vec{X}_0}(\vec{X})$ is the Dirac delta measure at $\vec{X}_0$. The probability density in (1.97) is an explicit solution of the Liouville equation in (1.96).

The empirical ensemble prediction in (1.97) is the central probability measure of interest for practical ensemble prediction. How can one quantify and estimate the uncertainty in such a prediction? In the remainder of this section, some of the important scientific issues associated with empirical ensemble prediction are discussed as well as the potential use of information theory in quantifying many aspects of uncertainty.

The first practical issue to confront is that the empirical ensemble size, R, in (1.97) is often very small in environmental science, on the order of $R \le 50$ for weather prediction and short term climate prediction in GCMs [**33, 44**] while, for example, $R = 500$ for intermediate models for predicting El Niño [**22, 23**]. This is due to the computational cost in these extremely complex systems in an operational setting requiring real time prediction. Furthermore, the idealized ensemble distribution, $p_0(\vec{X})$, is also unknown in practice. For theoretical studies in the perfect predictability scenario, the initial distribution, $p_0(\vec{X})$ from (1.96), is known exactly and the approximation by an ensemble $R \gg 1$ is the main emphasis. These studies in idealized systems are very useful in determining the features governing predictability in a given system [**19, 20**].

What is available to quantify the uncertainty from the empirical ensemble prediction p_t^E and readily evaluated are some of the moments,

$$\int_{\mathbb{R}^N} (\vec{X} - \bar{\vec{X}})^\alpha p_t^E(\vec{X})\, d\vec{X} = \frac{1}{R}\sum_{r=1}^{R} (\vec{X}_t^r - \bar{\vec{X}}^r)^\alpha, \quad 0 \le |\alpha| \le 2L, \tag{1.98}$$

with $\vec{X}^\alpha = X_1^{\alpha_1} \cdots X_N^{\alpha_N}$ [**18**], $|\alpha| = \sum \alpha_i$. Of course, in practice only the low order moments, where $L = 2$ or 4, are utilized for a judicious restricted set of variables

$$X_i, \quad 1 \le i \le M. \tag{1.99}$$

1.6.1. Practical and mathematical issues for predictability. Some of the main important issues in quantifying the uncertainty of an ensemble prediction are listed below:

(1.100)

A) How much lack of information is contained in a given prediction strategy, for example, measuring only the second moments in (1.98), compared with p_t^E?
B) When does the ensemble prediction distribution exhibit bimodality in some variable so that there are at least two different scenarios of significant change predicted in a given variable for times of interest? How can this be quantified in a computationally tractable fashion?
C) In a given dynamical system in (1.92), which subsets of variables, $X_1, \ldots, X_M$, are more predictable than the others? How can this be quantified?
D) For long term climate prediction, when is an ensemble prediction useful at all beyond the historical climate record? How can one estimate the lack of information in the climate record beyond the ensemble prediction? What features control the variation of this predictive utility in a given dynamical system?
E) What features characterize the rare ensemble predictions with more than typical information beyond the climate record?

All of the above issues involve the lack of information in one probability distribution, $\Pi(\vec{X})$, compared with another probability distribution, $p(\vec{X})$, over some subset of variables, $X_1, \ldots, X_M$. As discussed extensively in this chapter, in information theory this lack of information content is quantified by the relative entropy,

$$\mathcal{P}(p, \Pi) = \int_{\mathbb{R}^M} p \ln\left(\frac{p}{\Pi}\right). \tag{1.101}$$

(See [**10**]). Recently, in an important paper, Kleeman [**19**] has suggested the use of relative entropy to address the issue in (1.100) D) for long term climate prediction. In this case, Π is the historical record climate distribution on a given set of variables, while $p = p_t^E$ is the empirical ensemble prediction, so $\mathcal{P}(p_t^E, \Pi)$ should be estimated. The main issues in (1.100) D) were raised earlier by Anderson and Stern [**6**]. In the situation in (1.100) A), monitoring only the first and second moments of the prediction as is usually done in practice [**18, 44**] leads to the Gaussian prediction, p_G; so in this case $\mathcal{P}(p_t^E, p_G)$ should be estimated. Recently, Roulston and Smith [**37**] have suggested a similar use of information theory as a "score" or predictability measure for ensemble forecasts. Below we mention how to develop rigorous computationally feasible tests for bimodality to address (1.100) B) through non-Gaussian estimators for the empirical prediction ensemble. As regards (1.100) C), for the special case when both p and Π are Gaussian distributions, Schneider and Griffies [**40**] have developed the idea of principal predictability components utilizing the entropy difference; this interesting concept has been generalized by Majda, Kleeman, and Cai [**27**] to one typical practical situation where Π is Gaussian while p is non-Gaussian with (1.101) as the more precise measure of lack of information in Π. Finally, to address the important issue in (1.100) E) for long term prediction as well as the key factors that determine variability in predictive utility, Kleeman [**19**] has introduced and applied the concept of the signal/dispersion decomposition for the special case when both Π and p are Gaussian distributions. Recently, Kleeman, Majda, and Timofeyev [**21**] have determined the controlling factors of this facet of predictability in a simple model with statistical features of the atmosphere in a Gaussian setting, the truncated Burgers–Hopf model, earlier discussed in this chapter and also Chapter 2. This decomposition for suitable non-Gaussian measures, p and Π, has been developed by Majda, Kleeman and Cai in [**27**], and Abramov and Majda [**4**] developed a new, more precise, decomposition into a signal, dispersion, and cross-term. Both the signal and cross-term are directly and cheaply evaluated from the moments of p. Thus, they are readily available in determining variations in utility in a given dynamical system.

With the background discussion presented above, once one adopts the relative entropy in (1.101) as the theoretical tool for addressing the important issues in (1.100), the following practical issues arise:

(1.102)

A) How can one numerically compute or approximate the relative entropy, $P(p, \Pi)$, for a complex system with many degrees of freedom and a subset of variables with $M \gg 1$?

B) Are there computationally feasible and mathematically rigorous strategies to estimate $P(p, \Pi)$ from below, i.e. $P(p, \Pi) \geq P(p^*, \Pi)$, where $P(p^*, \Pi)$ can be evaluated either analytically or by a rapid numerical procedure?

To address these practical issues, Majda et al. [**27**] developed a more sophisticated framework of predictability through relative entropy for non-Gaussian probability density functions, which includes a hierarchy of rigorous lower bounds on relative entropy through the statistical moments beyond the mean and covariance through maximum entropy optimization. Abramov and Majda [**4**] converted the non-Gaussian predictability framework into a practical tool through the hierarchy of lower bounds and a rapid numerical optimization algorithm. Recently, Cai et al. [**9**] exhaustively tested several facets of the non-Gaussian information theoretic predictability framework in a simple chaotic mapping model with an explicit attractor ranging from Gaussian to fractal as parameters are varied. Kleeman and Majda [**20**] have quantified the loss of information in coarse-grained ensemble estimators and applied these ideas to geophysical turbulence. Different applications of relative entropy as a predictability tool were developed in Abramov and Majda [**4**]; besides, a straightforward measure of lack of information in the climate relative to the prediction ensemble, the relative entropy, can be used in estimating lack of information in a forecast ensemble relative to the actual events (Roulston and Smith [**37**]), in evaluating additional information content in the skewness and higher order forecast moments (non-Gaussianity), and the information flow between different subsets of phase space in an ensemble forecast (statistical correlation between different large scale phenomena). In Abramov and Majda [**4**] all of these facets were demonstrated for the Lorenz 96 model (Lorenz and Emanuel [**25**]), including highly non-Gaussian behavior. Finally, Abramov et al. [**5**] successfully applied the relative entropy framework to the simplest midlatitude atmospheric climate model, barotropic T21 spherical truncation with realistic orography in two dynamical regimes, mimicking the behavior of atmosphere at two different heights. In particular, the information flow was found responsible for correlated switches in large scale structures like the Arctic Oscillation, North Atlantic Oscillation, and Pacific/North American pattern. All of the above work demonstrates many practical facets of quantifying uncertainty in ensemble forecasts through the relative entropy; however, all of the work in idealized settings described above utilized large ensemble sizes. The work of Haven et al. [**16**] addresses the central practical issue of quantifying the uncertainty of an ensemble forecast with limited ensemble size in the non-Gaussian framework of information theory and relative entropy.

Perhaps, the most sophisticated contemporary uses of ensemble predictions in complex systems with small ensemble size occur in weather and climate predictions. Studies of predictions with forecast ensembles were performed, among others, by Anderson and Stern [**6**], Toth and Kalnay [**44**] and Palmer [**33**]. Practice shows that with current computational facilities, in order to perform real-time forecasts with large atmospheric GCMs, the size of the prediction ensemble has to be small, about 10–50 members, depending on spatial resolution, with the possibility of 50 to 100 member ensembles in the near future. Due to limited forecast ensemble size, certain complications arise concerning the credibility of information provided by such ensemble. The common strategy of dealing with small forecast ensembles is to maximize the information provided by the limited sample size, via generating a forecast ensemble in a very specific way. In particular, Ehrendorfer and Tribbia [**13**] show that for correct error growth reconstruction, the fastest growing directions of the phase space have to be sampled. Two efficient methods of ensemble generation are usually used in practical ensemble forecasts: local Lyapunov vectors (Toth and Kalnay [**44**], Kalnay [**18**]) and singular vectors (Palmer et al. [**34**], Reynolds and Palmer [**35**]). The efficiency of prediction depending on ensemble size has been studied previously by Buizza and Palmer [**8**]. The book of Kalnay [**18**] is strongly recommended for more reading on these topics.

In Section 2.3 and 2.4 of Chapter 2 we sketch new applications of information theory and the fluctuation-dissipation theorem to ensemble predictions near statistical equilibrium. A detailed introduction to utilizing information theory for ensemble predictions can be found in Chapter 15 of [**31**]. The interested reader is invited to look at the papers cited above for further details. This is an exciting practical topic which awaits further applications of mathematical ideas for complex chaotic nonlinear dynamical systems with many degrees of freedom.

References

1. R. Abramov, G. Kovačič, and A. J. Majda, *Hamiltonian structure and statistically relevant conserved quantities for the truncated Burgers–Hopf equation*, Comm. Pure Appl. Math. **56** (2003), 1–46.

2. R. Abramov and A. J. Majda, *Discrete approximations with additional conserved quantities: Deterministic and statistical behavior*, Methods Appl. Anal. **10** (2003), no. 2, 151–190.
3. ———, *Statistically relevant conserved quantities for truncated quasi-geostrophic flow*, Proc. Nat. Acad. Sci. U.S.A. **100** (2003), no. 7, 3841–3846.
4. ———, *Quantifying uncertainty for non-Gaussian ensembles in complex systems*, SIAM J. Sci. Comput. **26** (2004), no. 2, 411–447.
5. R. Abramov, A. J. Majda, and R Kleeman, *Information theory and predictability for low frequency variability*, J. Atmospheric Sci. (2004), in press.
6. J. Anderson and W. Stern, *Evaluating the potential predictive utility of ensemble forecasts*, Journal of Climate **9** (1996), 260–269.
7. C. Boucher, R. Ellis, and B. Turkington, *Derivation of maximum entropy principles in two-dimensional turbulence via large deviations*, J. Statist. Phys. **98** (2000), no. 5–6, 1235–1278.
8. R. Buizza and T. Palmer, *Impact of ensemble size on ensemble prediction*, Montly Weather Review **126** (1998), 2503–2518.
9. D. Cai, K. Haven, and A. J. Majda, *Quantifying predictability in a simple model with complex features*, Stoch. Dyn. **4** (2004), no. 4, 547–569.
10. T. Cover and J. Thomas, *Elements of information theory*, Wiley, New York, 1991.
11. M. DiBattista, A. Majda, and M.J. Grote, *Metastability of equilibrium statistical structures for prototype geophysical flows with damping and forcing*, Phys. D **151** (2001), no. 2–4, 271–304.
12. T. Dowling, *Dynamics of Jovian atmospheres*, Annu. Rev. Fluid Mech. **27** (1995), 293–334.
13. M. Ehrendorfer and J. Tribbia, *Optimal prediction of forecast error covariances through singular vectors*, J. Atmospheric Sci. **54** (1997), 286–313.
14. M. J. Grote and A. J. Majda, *Crude closure dynamics through large scale statistical theories*, Phys. Fluids **9** (1997), no. 11, 3431–3442.
15. ———, *Crude closure for flow with topography through large scale statistical theory*, Nonlinearity **13** (2000), no. 3, 569–600.
16. K. Haven, A. J. Majda, and R. Abramov, *Quantifying predictability through information theory: Small sample estimation in a non-Gaussian framework*, J. Comput. Phys. (2004), in press.
17. E. Jaynes, *Information theory and statistical mechanics*, Phys. Rev. **106** (1957), 620–630.
18. E. Kalnay, *Atmospheric modeling, data assimilation and predictability*, Cambridge University Press, New York, 2003.
19. R. Kleeman, *Measuring dynamical prediction utility using relative entropy*, J. Atmospheric Sci. **59** (2002), 2057–2072.
20. R. Kleeman and A. J. Majda, *Predictability in a model of geophysical turbulence*, J. Atmospheric Sci. (2004), In press.
21. R. Kleeman, A. J. Majda, and I. Timofeyev, *Quantifying predictability in a model with statistical features of the atmosphere*, Proc. Nat. Acad. Sci. **99** (2002), 15291–15296.
22. R. Kleeman and A. Moore, *A theory for the limitation of ENSO predictability due to stochastic atmospheric transients*, J. Atmospheric Sci. **54** (1997), 753–767.
23. ———, *A new method for determining the reliability of dynamical ENSO predictions*, Montly Weather Review **127** (1999), 694–705.
24. J. Lamperti, *Probability*, Benjamin Inc., New York, 1966.
25. E. Lorenz and K. Emanuel, *Optimal sites for supplementary weather observations*, J. Atmospheric Sci. **55** (1998), 399–414.
26. A. J. Majda and A. Bertozzi, *Vorticity and incompressible flow*, Cambridge University Press, Cambridge, England, 2001.
27. A. J. Majda, R. Kleeman, and D. Cai, *A framework for predictability through relative entropy*, Methods Appl. Anal. **9** (2002), 425–444.
28. A. J. Majda and I. Timofeyev, *Remarkable statistical behavior for truncated Burgers-Hopf dynamics*, Proc. Nat. Acad. Sci. **97** (2000), no. 23, 12413–12417.
29. ———, *Statistical mechanics for truncations of the Burgers–Hopf equation: A model for intrinsic stochastic behavior with scaling*, Milan J. Math. **70** (2002), 39–96.
30. A. J. Majda and X. Wang, *The emergence of large scale coherent structure under small scale random bombardments*, Comm. Pure Appl. Math. (2004), in press.
31. ———, *Nonlinear dynamics and statistical theories for basic geophysical flows*, Cambridge University Press, New York, 2005.

32. P. Marcus, *Jupiter's Great Red Spot and other vortices*, Annu. Rev. Astrophys. **31** (1993), 523–573.
33. T. Palmer, *Predicting uncertainty in forecasts of weather and climate*, Rep. Progr. Phys. **63** (2000), 71–116.
34. T. Palmer, F. Molteni, R. Mureau, R. Buizza, P. Chapelet, and J. Tribbia, *Ensemble prediction*, Proc. Validation of Models over Europe, vol. 1, 1993, pp. 21–66.
35. C. Reynolds and T. Palmer, *Decaying singular vectors and their impact on analysis and forecast correction*, J. Atmospheric Sci. **55** (1998), 2576–2596.
36. J. Rogers, *The giant planet Jupiter*, Cambridge University Press, 1995.
37. M. Roulston and L. Smith, *Evaluating probabilistic forecasts using information theory*, Monthly Weather Review **130** (2002), 1653–1660.
38. R. Salmon, *Lectures on geophysical fluid dynamics*, Oxford Press, New York, 1998.
39. R. Salmon, G. Holloway, and M. Hendershott, *The equilibrium statistical mechanics of simple quasi-geostrophic models*, J. Fluid Mech. **75** (1976), 691–703.
40. T. Schneider and S. Griffies, *A conceptual framework for predictability studies*, Journal of Climate **12** (1999), 3133–3155.
41. C. Shannon, *The mathematical theory of communication*, Bell System Tech. J. (1948).
42. L. Tartar, *Compensated compactness and applications to partial differential equations*, Nonlinear Analysis and Mechanics: Heriot–Watt Symposium (Pitman, Boston), vol. 39, Research Notes in Mathematics Series, 1979, pp. 136–212.
43. C. Thompson, *Mathematical statistical mechanics*, Princeton University Press, Princeton, New Jersey, 1972.
44. Z. Toth and E. Kalnay, *Ensemble forecasting at* NMC: *The generation of perturbations*, Bull. Amer. Meteorol. Soc. **74** (1993), 2317–2330.
45. B. Turkington, A. Majda, K. Haven, and M. DiBattista, *Statistical equilibrium predictions of jets and spots on Jupiter*, Proc. Nat. Acad. Sci. **98** (2001), 12346–12350.
46. V. Zeitlin, *Finite-mode analogues of 2D ideal hydrodynamics: Coadjoint orbits and local canonical structure*, Phys. D **49** (1991), 353–362.

CHAPTER 2

The Fluctuation-Dissipation Theorem for Complex Nonlinear Systems

2.1. Introduction

The fluctuation-dissipation theorem (FDT) is one of the cornerstones of modern statistical physics [**6, 12**]. Roughly speaking, this theorem states that for systems in statistical equilibrium the average mean response to small external perturbations can be calculated through the knowledge of suitable correlation functions of the unperturbed statistical system. This theorem has had great practical use in traditional settings involving the statistical equilibrium of baths of identical gas or liquid molecules [**6, 12**].

Despite the fact that the climate system is a complex chaotic multiscale system with forcing and dissipation, in the mid 1970s Leith [**13**] suggested the possibility that the fluctuation-dissipation theorem might be valid for appropriate variables in the complex climate system. If this can be established, such a result has profound implications for studying climate change scenarios. Namely, the response of the present climate to small changes in the external forcing, at least for these appropriate variables, can be calculated through statistics of the current climate alone. Leith's suggestion has inspired others such as Bell [**2**], Carnevale et al. [**3**], and Dymnikov, Gritsoun and Branstator [**8, 9, 10**] to study the potential validity of the fluctuation-dissipation theorem for idealized climate models with various approximations and numerical procedures. In another direction, Craig and Cohen [**4**] have attempted to apply these ideas to the unresolved features of tropical convection. While many of the results in this work are encouraging, no systematic understanding of the validity of various approximations to the fluctuation-dissipation theorem has been achieved. Given the practical importance as well as the interesting combination of mathematical and physical ideas involved, the validity of the fluctuation-dissipation theorem for complex multiscale nonlinear systems is obviously a topic of great importance for future research using modern applied mathematics.

In this chapter we present a comprehensive introduction to the fluctuation-dissipation theorem for complex nonlinear systems together with new results on information flow with perturbed forcing and also for the behavior of ensemble predictions near equilibrium. We also present new and comprehensive results testing the validity of the predictions of FDT for three prototype chaotic equilibrium systems with diverse statistical behavior which are utilized as toy models for various features of the atmosphere, the truncated Burgers–Hopf model (TBH), the Kruskal–Zabusky (K-Z) finite-difference scheme, and the Lorenz 96 model without forcing and dissipation (IL96). These three models have the same equipartitioned equilibrium climate distribution but exhibit diverse multiple timescales with strongly chaotic dynamics in statistical equilibrium. Here the validity of the FDT approximation is studied for the full nonlinear operator for all state space variables for a wide range of response times in each of the three examples. In this context, the predictions of FDT are expected to be exact so that various numerical issues and sensitivity to detailed nonlinear dynamics can be explored; for these idealized complex nonlinear systems with multiple time scales but in statistical equilibrium the predictions of FDT are robust and excellent. A crucial issue in future applications of FDT to climate modeling involves its potential use for complex nonlinear systems with dissipation and forcing. All of the applications of FDT developed so far in this context [**2, 3, 10**] involve what we call below the "quasi-Gaussian approximation". In this chapter we give new mathematical insights which help to explain the validity of such approximations for the special functionals for the FDT response used so far in practice. We also give a new comprehensive application of this approximation for FDT to the damped forced Lorenz 96 model for a wide range of forcing and also provide direct links to the statistical dynamics of IL96 in a novel fashion. Finally, we briefly discuss future research directions for improving and testing FDT for complex chaotic dynamical systems.

2.1.1. Preliminaries on the fluctuation-dissipation theorem. In this chapter, to make matters concrete, we discuss FDT for an unperturbed system given by a large system of stochastic ODEs for a vector $\vec{X} \in \mathbb{R}^N$, $N \gg 1$, given by

$$d\vec{X} = \vec{F}(\vec{X})\,dt + g(\vec{X})\,d\vec{W}, \tag{2.1}$$

where $\vec{F}$ is a vector field of dimension N, $\vec{W}$ is a standard K-dimensional Wiener process, and g is a $N \times K$ matrix valued function. As discussed earlier in Chapter 1, in the statistical dynamics of complex systems we are interested in the evolution of probability densities, $p_t(\vec{X}) = p(\vec{X}, t)$, associated with (2.1) rather than individual solutions. It is well-known that the evolution of these probability densities, $p_t(\vec{X})$, satisfies the Fokker–Planck equation [**20**]

$$\begin{aligned} \frac{\partial}{\partial t} p_t(\vec{X}) &= -\operatorname{div}_{\vec{X}}(\vec{F} p_t) + \sum_{ij} \frac{\partial^2}{\partial X_i \partial X_j}\left(\frac{Q_{ij}}{2} p_t\right), \\ p_t(\vec{X})\big|_{t=0} &= p_0(\vec{X}), \end{aligned} \tag{2.2}$$

where $Q(\vec{X})$ is the $N \times N$ matrix, $Q = g(\vec{X})g(\vec{X})^T$. We call the linear operator on the right hand side of (2.2) $L_{\mathrm{FP}}(\vec{X})$, so that (2.2) becomes

$$\begin{aligned} \frac{\partial p_t}{\partial t} &= L_{\mathrm{FP}} p_t, \\ p_t\big|_{t=0} &= p_0. \end{aligned} \tag{2.3}$$

In the special case where the external noise vanishes, so that $Q \equiv 0$, the Fokker–Planck operator in (2.3) reduces to the Liouville operator,

$$\frac{\partial p}{\partial t} = L_{\mathrm{FP}} p = -\operatorname{div}_{\vec{X}}(\vec{F} p). \tag{2.4}$$

Now let us perturb the dynamical system in (2.1) by a small external forcing, $\delta\vec{F}(\vec{X}, t)$,

$$d\vec{X} = \left(\vec{F}(\vec{X}) + \delta\vec{F}(\vec{X}, t)\right)dt + g(\vec{X})\,d\vec{W}. \tag{2.5}$$

Then, as in (2.2) and (2.3), the perturbed probability density $p^\delta(\vec{X},t) = p_t^\delta(\vec{X})$ satisfies the perturbed Fokker–Planck equation,

$$\begin{aligned} \frac{\partial}{\partial t} p^\delta(\vec{X},t) &= L_{\mathrm{FP}} p^\delta(\vec{X},t) + \delta L_{\mathrm{ext}} p^\delta(\vec{X},t), \\ p^\delta(\vec{X},0) &= p_0^\delta(\vec{X}), \end{aligned} \tag{2.6}$$

with

$$\delta L_{\mathrm{ext}} p = -\operatorname{div}_{\vec{X}} \left(\delta \vec{F}(\vec{X},t) p\right). \tag{2.7}$$

Now let $p_{\mathrm{eq}}(\vec{X})$ be smooth rapidly decreasing equilibrium probability measure for the unperturbed system in (2.2) so that

$$L_{\mathrm{FP}} p_{\mathrm{eq}} = 0. \tag{2.8}$$

The expected value of a nonlinear functional, $A(\vec{X})$, with respect to this equilibrium measure is given by

$$E[A(\vec{X})] = \int_{\mathbb{R}^N} A(\vec{X}) p_{\mathrm{eq}}(\vec{X})\, d\vec{X}. \tag{2.9}$$

Similarly, for the perturbed system in (2.5) the expected value of the nonlinear functional $A(\vec{X})$ at time t is

$$E^\delta[A(\vec{X})] = \int_{\mathbb{R}^N} A(\vec{X}) p^\delta(\vec{X},t)\, d\vec{X}. \tag{2.10}$$

The fluctuation-dissipation theorem provides a formal procedure to calculate the change in the expected value

$$\delta E[A(\vec{X})](t) = E^\delta[A(\vec{X})] - E[A(\vec{X})] \tag{2.11}$$

through knowledge of suitable functionals of the unperturbed system in (2.9) for $\delta \ll 1$. Of course, the tacit assumption that the initial probability density is close to the unperturbed equilibrium also needs to be made, i.e.

$$p_0^\delta(\vec{X}) = p_{\mathrm{eq}}(\vec{X}) + \delta p_0(\vec{X}) \tag{2.12}$$

with $\delta \ll 1$. In the derivation below we assume for simplicity that $\delta p_0(\vec{X}) \equiv 0$. All calculations in the remainder of this chapter will involve concise mathematical reasoning but no rigorous justification will be supplied including the stated theorems; as already used in (2.9) and (2.10) tacit assumptions of smoothness and decay for large values of $\|\vec{X}\|$ for $p_{\mathrm{eq}}(\vec{X})$ will be imposed as required.

2.2. Derivation of the fluctuation-dissipation theorem

Our derivation below follows that in Risken [**20**]. First, we assume a very natural explicit time-separable structure for the perturbation $\delta\vec{F}(\vec{X},t)$ in (2.5), namely

$$\delta F_i(\vec{X},t) = a_i(\vec{X}) \delta F_i(t), \quad 1 \le i \le N, \tag{2.13}$$

so that δL_{ext} in (2.7) can be written concisely as

$$\delta L_{\mathrm{ext}} p = \vec{L}_{\mathrm{ext}} p \cdot \delta \vec{F}(t) \tag{2.14}$$

with the row vector $\vec{L}_{\mathrm{ext}}$ given by

$$\vec{L}_{\mathrm{ext}} p = -\frac{\partial}{\partial X_i}\left(\vec{a}_i(\vec{X}) p\right), \quad 1 \le i \le N. \tag{2.15}$$

We write $p^\delta(\vec{X},t)$ from (2.6) in the form

$$p^\delta(\vec{X},t) = p_{\text{eq}}(\vec{X}) + \delta p(\vec{X},t), \tag{2.16}$$

where δp is a small perturbation consistent with the fact that δ in (2.13) is a small parameter in the external perturbation. By using $L_{\text{FP}} p_{\text{eq}} = 0$ from (2.8) together with (2.6), we obtain the exact equation for $\delta p(\vec{X},t)$

$$\frac{\partial}{\partial t}\delta p(\vec{X},t) = L_{\text{FP}}\delta p + \delta L_{\text{ext}} p_{\text{eq}} + \delta L_{\text{ext}}\delta p. \tag{2.17}$$

The last term in (2.17) is formally $\mathcal{O}(\delta^2)$ for small values of δ. The first step in the FDT derivation is to ignore this quadratic term and arrive at a linear equation for δp,

$$\begin{aligned} \frac{\partial}{\partial t}\delta p(\vec{X},t) &= L_{\text{FP}}\delta p + \delta L_{\text{ext}} p_{\text{eq}}, \\ \delta p\big|_{t=0} &= 0. \end{aligned} \tag{2.18}$$

Since L_{FP} is a linear operator, with the semigroup notation, $\exp[tL_{\text{FP}}]$, for this solution operator, the solution of (2.18) is written concisely as

$$\delta p = \int_0^t \exp[(t-t')L_{\text{FP}}]\big(\delta L_{\text{ext}}(t')p_{\text{eq}}\big)\,dt'. \tag{2.19}$$

Now, combining (2.19) with (2.11), (2.13), (2.14) and (2.16), we have the *linear response formula*

$$\delta E[A(\vec{X})](t) = \int_{\mathbb{R}^N} A(\vec{X})\delta p(\vec{X},t)\,d\vec{X} = \int_0^t \vec{R}(t-t')\cdot\delta\vec{F}(t')\,dt', \tag{2.20}$$

where the vector linear response operator is given by

$$\vec{R}(t) = \int_{\mathbb{R}^N} A(\vec{X})\big(\exp[tL_{\text{FP}}][\vec{L}_{\text{ext}} p_{\text{eq}}]\big)(\vec{X})\,d\vec{X}. \tag{2.21}$$

This general calculation is the first step in the fluctuation-dissipation theorem. For nonlinear systems with many degrees of freedom, direct use of the formula in (2.21) is completely impractical because the exponential, $\exp[tL_{\text{FP}}]$, can not be calculated directly. However, the fluctuation-dissipation theorem states that

THEOREM 2.1. *The computation of the linear response vector $\vec{R}(t)$ in (2.21) can be reduced to the computation of the statistical correlation vector*

$$\vec{R}(\tau) = \langle A(\vec{X}(t+\tau))\vec{B}(\vec{X}(t))\rangle, \tag{2.22}$$

with the special nonlinear functional,

$$\vec{B}(\vec{X}) \equiv \frac{\vec{L}_{\text{ext}}(\vec{X})p_{\text{eq}}(\vec{x})}{p_{\text{eq}}(\vec{x})}, \tag{2.23}$$

where the correlation vector on the right hand size of (2.22) is evaluated at the unperturbed state and defined below in (2.24).

PROOF. We write the correlation function for two arbitrary functionals $A(\vec{X})$ and $B(\vec{X})$ as

$$\langle A(\vec{X}(t+\tau))B(\vec{X}(t))\rangle = \int_{\mathbb{R}^N} A(\vec{X})\int_{\mathbb{R}^N} B(\vec{Y})p(\vec{X},t+\tau;\vec{Y},t)\,d\vec{Y}\,d\vec{X}, \tag{2.24}$$

where $p(\vec{X}, t_1; \vec{Y}, t_2)$ is the joint probability density function for the unperturbed system in (2.1). In the statistical steady state $p(\vec{X}, t+\tau; \vec{Y}, t)$ depends only on $\vec{X}$, $\vec{Y}$ and τ, and can be written as a product of a transition (or conditional) probability density $p_{\text{cond}}(\vec{X}, t \mid \vec{Y}, 0)$ and the invariant probability density $p_{\text{eq}}(\vec{Y})$ from (2.8) for the unperturbed system in (2.1):

$$p(\vec{X}, t+\tau; \vec{Y}, t) = p(\vec{X}, \tau; \vec{Y}, 0) = p_{\text{cond}}(\vec{X}, \tau \mid \vec{Y}, 0) p_{\text{eq}}(\vec{Y}). \tag{2.25}$$

The time evolution of the transition probability density itself obeys the unperturbed Fokker–Planck equation

$$\begin{aligned} \frac{\partial}{\partial t} p_{\text{cond}}(\vec{X}, t \mid \vec{Y}, 0) &= L_{\text{FP}}(\vec{X}) p_{\text{cond}}(\vec{X}, t \mid \vec{Y}, 0), \\ p_{\text{cond}}(\vec{X}, 0 \mid \vec{Y}, 0) &= \delta(\vec{X} - \vec{Y}) \end{aligned} \tag{2.26}$$

with formal solution

$$p_{\text{cond}}(\vec{X}, t \mid \vec{Y}, 0) = \exp[t L_{\text{FP}}(\vec{X})] \delta(\vec{X} - \vec{Y}). \tag{2.27}$$

Substituting (2.25) and (2.27) into (2.24) yields

$$\langle A(\vec{X}(t+\tau)) B(\vec{X}(t)) \rangle = \int_{\mathbb{R}^N} A(\vec{X}) \exp[\tau L_{\text{FP}}(\vec{X})] \big[B(\vec{X}) p_{\text{eq}}(\vec{X})\big] \, d\vec{X}. \tag{2.28}$$

Now it is obvious that the linear response vector $\vec{R}(\tau)$ in (2.22) coincides with the correlation function $\langle A(\vec{X}(t+\tau)) B(\vec{X}(t)) \rangle$ from (2.28) if the relation (2.23) holds. This concludes the proof. □

Note that the correlation function can be evaluated from historical observations without explicit knowledge of the right hand side of the unperturbed system in (2.1).

2.2.1. Complex nonlinear models. There is a wide class of nonlinear systems with complex multiscale features where the FDT just presented above has special interest. These models are the special case of (2.1) where

$$\frac{d\vec{X}}{dt} = L\vec{X} + \mathcal{B}(\vec{X}, \vec{X}) \equiv \vec{F}(\vec{X}) \tag{2.29}$$

with L a skew-symmetric linear operator and $\mathcal{B}(\vec{X}, \vec{X})$ a bilinear operator satisfying

(2.30)
- the Liouville property, $\operatorname{div}_{\vec{X}} \mathcal{B}(\vec{X}, \vec{X}) = 0$,
- the conservation condition, $\vec{X} \cdot \mathcal{B}(\vec{X}, \vec{X}) = 0$.

As discussed in Chapter 1, as a consequence of (2.30) the equations in (2.29) conserve the energy, $E = \frac{1}{2}\vec{X} \cdot \vec{X} = \frac{1}{2}\|\vec{X}\|^2$; also, with (2.30) and the fact that L is skew-symmetric, the Liouville property is satisfied for the equation in (2.4) so that $\operatorname{div}_{\vec{X}} \vec{F} = 0$. These two facts together imply that there is a family of Gaussian invariant probability measures for the unperturbed system, $p_{\text{eq}}(\vec{X})$, satisfying (2.8) with

$$p_{\text{eq}}(\vec{X}) = C_\theta \exp\big(-\theta E(\vec{X})\big) = C_\theta \exp\Big(-\frac{\theta}{2}\vec{X} \cdot \vec{X}\Big). \tag{2.31}$$

In (2.31) C_θ is just a normalizing constant to guarantee a probability measure while $\theta > 0$ is a normalizing constant describing the mean energy in the system. Nontrivial examples with the structure in (2.29)–(2.31) are prevalent for the large scale structure of simplified climate models such as truncated geophysical flow in pseudoenergy variables; see Chapter 8 of [**18**] as well as Section 6 of [**17**]. Three other interesting examples for studying the validity of FDT for systems with the structure in (2.29)–(2.31) include

- The TBH model [**15**]
$$(u_\Lambda)_t = -\tfrac{1}{2}P_\Lambda(u_\Lambda^2)_x,$$
$$P_\Lambda f = \sum_{|k|\le J/2} \hat{f}_k e^{ikx}, \quad \bar{\hat{f}}_{-k} = \hat{f}_k.$$
- The K-Z model [**16**]
$$\frac{du_j}{dt} = -\frac{F_{j+1/2} - F_{j-1/2}}{\Delta x}, \quad j = 0, 1, \dots, J \tag{2.32}$$
with $F_{j+1/2} = \frac{1}{6}(u_j^2 + u_j u_{j+1} + u_{j+1}^2)$ and periodic boundary conditions, $u_0 = u_J$.
- The IL96 model [**1**]
$$\frac{du_j}{dt} = (u_{j+1} - u_{j-2})u_{j-1}, \quad j = 0, 1, \dots, J, \qquad u_0 = u_J.$$

These are elementary toy models with analogues for some of the statistical features of the atmosphere. Detailed verification of the properties in (2.29)–(2.31) for the three systems in (2.32) can be found in Chapter 7 of [**18**] as well as the references noted above. In each of the three systems in (2.30) the variables $\vec{X}$ can be identified with either the physical coordinates, $X_j = u_j$, or the Fourier coefficient variables, $X_j = a_j$, b_j where $\hat{u}_k = a_k + ib_k$, $0 \le k \le J$ are the Fourier coefficients for non-negative values of k. All of this follows readily from the isometry between discrete periodic lattice values and finite Fourier coefficients induced by the discrete Fourier transform (see [**16**]). Below we use physical space and Fourier space interchangeably in discussing statistical features and the validity of FDT. See Section 2.8 for a similar use of the discrete Fourier transform.

2.2.2. Predictions of FDT for complex nonlinear models. The simplest and important perturbations of a nonlinear system as in (2.29) for climate modeling involve the response to external forcing. This is the climate change problem in the above models. In this special case,

$$\vec{L}_{\text{ext}} = -\frac{\partial}{\partial X_i}, \quad 1 \le i \le N. \tag{2.33}$$

For these special perturbations due to external forcing, the Theorem 2.1 states that the linear response operator $\vec{R}(\tau)$ needed to compute $\delta E[A(\vec{X})](t)$ in (2.20) is given by the formula

$$\vec{R}(\tau) = \theta\big\langle A\big(\vec{X}(t+\tau)\vec{X}(t)\big)\big\rangle. \tag{2.34}$$

The simplest important special case of (2.34) arises when we are interested in computing the mean response of the state vectors, $\vec{X}$, themselves to changes in external forcing: thus, we use $A(\vec{X}) = X_i$, $1 \le i \le N$ and the Theorem 2.1 states that the $N \times N$ linear response matrix needed for $\delta E[X_i](t)$ is given by

$$\big(R_{ij}(\tau)\big) = \theta\langle \vec{X}(t+\tau) \otimes \vec{X}(t)\rangle, \tag{2.35}$$

i.e. the correlation matrix for the unperturbed system. The tests of FDT for climate modeling until now have only utilized the special functionals, $A(\vec{X}) = X_i$ (see [**2, 3, 8, 9, 10**]). However, we mention here that other nonlinear functionals such as $A(\vec{X}) = X_i^2$, $X_i X_j$ are certainly important in practical applications for estimating the fluctuations in the perturbed climate. We demonstrate this subsequently in this chapter in the context of ensemble prediction.

Also, there are other meaningful external perturbations besides external forcing. For example, adding the perturbed dissipation, $-\delta \bar{d} X_i$, to (2.1) induces the operator

$$\vec{L}_{\text{ext}} p = -\bar{d} \frac{\partial}{\partial X_i}(X_i p), \quad 1 \leq i \leq N, \tag{2.36}$$

so that in Theorem 2.1 $\vec{B} = (B_i)$, $1 \leq i \leq N$ with

$$B_i = \frac{L^i_{\text{ext}} p_{\text{eq}}}{p_{\text{eq}}} = \bar{d}(-1 + \theta X_i^2). \tag{2.37}$$

In Sections 2.6 and 2.9 of this chapter we test the prediction of Theorem 2.1 for the three ideal models in (2.32) which exhibit widely varying changes in behavior of the response operator with varying external forcing.

2.3. Information flow in the response to small external perturbations

A basic question to ask in climate change scenarios is the following one: which small external forcing, $\delta \vec{F}(t)$, create the most additional information relative to the climatology? The external forcing directions with the largest additional information are clearly the most significant ones for climate change. In Chapter 1 we discussed that the lack of information in a prior probability measure Π relative to a given probability measure p is given by the relative entropy,

$$\mathcal{P}(p, \Pi) = \int_{\mathbb{R}^N} p \ln\left(\frac{p}{\Pi}\right). \tag{2.38}$$

Thus, a quantitative way to address the information flow in the above climate change scenario is to measure

$$\mathcal{P}\big(p^\delta(t), p_{\text{eq}}\big) \tag{2.39}$$

as time evolves, where $p^\delta(\vec{X}, t)$ satisfies the perturbed Fokker–Planck equation in (2.6). With the decomposition $p^\delta(\vec{X}, t) = \delta p + p_{\text{eq}}(\vec{X})$, familiar from our discussion of FDT, we write

$$\begin{aligned}
\mathcal{P}(p^\delta, p_{\text{eq}}) &= \int_{\mathbb{R}^N} (p_{\text{eq}} + \delta p) \ln\left(1 + \frac{\delta p}{p_{\text{eq}}}\right) \\
&= \int_{\mathbb{R}^N} (p_{\text{eq}} + \delta p)\left(\frac{\delta p}{p_{\text{eq}}} - \frac{1}{2}\left(\frac{\delta p}{p_{\text{eq}}}\right)^2\right) + \mathcal{O}(\|\delta p\|^3) \\
&= \int_{\mathbb{R}^N} \delta p + \frac{1}{2}\int_{\mathbb{R}^N} \frac{\delta p^2}{p_{\text{eq}}} + \mathcal{O}(\|\delta p\|^3) \\
&= \frac{1}{2}\int_{\mathbb{R}^N} \frac{\delta p^2}{p_{\text{eq}}} + \mathcal{O}(\|\delta p\|^3) = \tilde{\mathcal{P}}(p^\delta, p_{\text{eq}}) + \mathcal{O}(\|\delta p\|^3),
\end{aligned} \tag{2.40}$$

In (2.40) we utilized the fact that $\delta p(\vec{X},t)$ is the difference between two probability densities and therefore its spatial integral is zero. By $\tilde{P}\big(p^\delta(\vec{X},t), p_{\rm eq}(\vec{X})\big)$ we further denote the leading order term in the expansion of the relative entropy in (2.40):

$$\tilde{P}\big(p^\delta(\vec{X},t), p_{\rm eq}(\vec{X})\big) = \frac{1}{2}\int_{\mathbb{R}^N} \frac{\delta p(\vec{X},t)^2}{p_{\rm eq}(\vec{X})}\, d\vec{X}. \tag{2.41}$$

In Theorem 2.1 we have assumed that $p^\delta(\vec{X},t)$ is differentiable at $\delta = 0$. Continuing with this smoothness assumption, we have

$$p^\delta = p_{\rm eq} + \delta \frac{dp^\delta}{d\delta} + o(\delta),$$

so that

$$\tilde{P}\big(p^\delta(t), p_{\rm eq}\big) = \frac{1}{2}\delta^2 \int_{\mathbb{R}^N} \frac{(dp^\delta/d\delta)^2}{p_{\rm eq}}\, d\vec{X} = \delta^2 I\big(p^\delta(t)\big) \tag{2.42}$$

where $I\big(p^\delta(t)\big)$ is the Fisher information of the family of probability measures, $p^\delta(t)$, evaluated at $\delta = 0$ [**23**]. Thus, $\tilde{P}$ is a multiple of Fisher information.

There is a beautiful formula for the evolution of the leading order information flow $\tilde{P}\big(p^\delta(t), p_{\rm eq}\big)$ as time evolves provided that the dynamical system in (2.1) has no external noise. In fact, we have the following

THEOREM 2.2. *Assume that the nonlinear equations in* (2.1) *have no external noise, i.e.* $g \equiv 0$, *and the Liouville equation for the vector field* $\vec{F}$ *in* (2.4) *has a smooth nonvanishing equilibrium density* $p_{\rm eq}$. *Then the leading order information flow* $\tilde{P}(p^\delta(t), p_{\rm eq})$ *defined in* (2.40) *and* (2.41) *satisfies the identity*

$$\frac{d\tilde{P}}{dt} = \delta E\left[\frac{\vec{L}_{\rm ext}(\vec{X})p_{\rm eq}}{p_{\rm eq}}\right]\cdot \delta\vec{F}. \tag{2.43}$$

Thus, by integrating (2.43) *in time and computing* δE *by Theorem* 2.1 *from Section* 2.2, *we obtain*

$$\tilde{P}\big(p^\delta(T), p_{\rm eq}\big) = \int_0^T dt \int_{-\infty}^t dt'\delta\vec{F}(t)\cdot R(t-t')\cdot\delta\vec{F}(t') \tag{2.44}$$

where $R(\tau)$ *is the correlation matrix,*

$$R(\tau) = \left\langle \frac{\vec{L}_{\rm ext}p_{\rm eq}}{p_{\rm eq}}\Big|_{\vec{X}(t+\tau)} \otimes \frac{\vec{L}_{\rm ext}p_{\rm eq}}{p_{\rm eq}}\Big|_{\vec{X}(t)}\right\rangle. \tag{2.45}$$

Recall that $\vec{L}_{\rm ext}p$ *is defined above explicitly in* (2.14) *and* (2.15).

Note that for the general complex nonlinear models discussed in (2.32) with external forcing from (2.29), we calculate from Theorem 2.2 that $R(\tau)$ in (2.45) is given by

$$R(\tau) = \theta^2\langle \vec{X}(t+\tau)\otimes\vec{X}(t)\rangle. \tag{2.46}$$

Thus, the leading order information flow in response to small external forcing is determined by the properties of the integrated correlation matrix. Of course, physical and mathematical consistency requires that since $\tilde{P}\big(p^\delta(t), p_{\rm eq}\big) \ge 0$, the right hand side of (2.44) must be nonnegative in general. To see this, note that the correlation matrix $R(\tau)$ given in (2.45) is the correlation matrix of the random variable

$\vec{B} = \vec{L}_{\text{ext}} p_{\text{eq}} / p_{\text{eq}}$; according to the Bochner–Khinchin Theorem, the Fourier transform matrix in time, $\hat{R}(\omega)$, is a nonnegative matrix for each frequency, ω. For $\delta\vec{F}(t)$ belonging to $C^\infty(0,T)$, Plancherel's formula for the time-Fourier transform of the right hand side of (2.44) yields that this term equals

$$\int_{-\infty}^{\infty} \delta\hat{\vec{F}}(\omega) \cdot \hat{R}(\omega) \cdot \delta\bar{\hat{\vec{F}}}(\omega)\, d\omega \geq 0$$

which is nonnegative as required since $\hat{R}(\omega)$ is a nonnegative matrix. A simple approximation argument allows us to conclude the result for more general $\delta\vec{F}(t)$.

Let us return to the climate change scenario which began our discussion in this section and attempt to find the constant external forcing directions associated with $\vec{L}_{\text{ext}}$ which induce the largest information flow in a time, T, for small external perturbations as in (2.13)–(2.15). Thus, assuming that $\delta\vec{F}(t) = \delta\vec{F}_0$ is constant in time for $0 \leq t \leq T$, we use Theorem 2.2 and compute

$$\begin{aligned} \tilde{P}(p^\delta(t), p_{\text{eq}}) &= \int_0^T \int_0^t \delta\vec{F}_0 \cdot R(t-t') \cdot \delta\vec{F}_0\, dt'\, dt \\ &= \delta\vec{F}_0 \cdot \int_0^T \int_0^t R_{\text{sym}}(t-t')\, dt'\, dt \cdot \delta\vec{F}_0 \qquad (2.47) \\ &= \delta\vec{F}_0 \cdot \int_0^T (T-s) R_{\text{sym}}(s)\, ds \cdot \delta\vec{F}_0. \end{aligned}$$

From the first equality in (2.47) it is clear that only the symmetric part $R_{\text{sym}} = \frac{1}{2}(R + R^T)$ of R enters the above formula; furthermore, our discussion in the preceding paragraph establishes that the matrix

$$M(T) = \frac{1}{T} \int_0^T (T-s) R_{\text{sym}}(s)\, ds \qquad (2.48)$$

is a positive definite symmetric matrix. Now, it is easy to solve the problem posed earlier. Clearly, the subspace identified with the largest eigenvalue of $M(T)$ is the direction of external forcing, $\delta\vec{F}_0$, with the largest information content in the response at time T; a hierarchy of response directions is calculated by utilizing all the eigenspaces of $M(T)$ with their corresponding nonnegative eigenvalues.

If the dynamical system in (2.1) is strongly mixing in the sense that for the special functions $\vec{B} = \vec{L}_{\text{ext}} p_{\text{eq}} / p_{\text{eq}}$ the correlation matrix $R(\tau)$ in (2.45) decays sufficiently rapidly in time, the matrix $M(T)$, defined in (2.48) has a limit as $T \to \infty$ given by

$$M(\infty) = \int_0^\infty R_{\text{sym}}(s)\, ds.$$

Thus, the natural low frequency basis for long times and constant external forcing is the one which diagonalizes the symmetric matrix $M(\infty)$. Through very different considerations, Dymnikov and Gritsoun **[9]** have introduced this optimal low frequency basis for calculating the FDT response; here, we have shown that such a basis arises naturally by decomposing the dynamical system into a hierarchy of directions according to the strength of the information flow in response to external perturbations.

We still owe the reader the proof of Theorem 2.2. Here is the proof. Using (2.6), the time evolution equation for $P[p^\delta(t), p_{\text{eq}}]$ is

$$\frac{dP}{dt} = \int_{\mathbb{R}^N} \ln\left(1 + \frac{\delta p}{p_{\text{eq}}}\right) L_{\text{FP}}\delta p + \int_{\mathbb{R}^N} \ln\left(1 + \frac{\delta p}{p_{\text{eq}}}\right) \delta L_{\text{ext}} p_{\text{eq}}. \tag{2.49}$$

The equation in (2.49), truncated to leading order, becomes the exact evolution equation for the leading order relative entropy term in (2.41):

$$\frac{d\tilde{P}}{dt} = \int_{\mathbb{R}^N} \frac{\delta p}{p_{\text{eq}}} L_{\text{FP}}(\vec{X})\delta p + \int_{\mathbb{R}^N} \frac{\delta p}{p_{\text{eq}}} \delta L_{\text{ext}} p. \tag{2.50}$$

The above statement can be directly verified by differentiating (2.41) in time and using the linearization in (2.18) as a substitute for the time derivative of $\delta p(\vec{X}, t)$. By $L_L(\vec{X})$ we denote a general Fokker–Planck operator for the deterministic dynamical systems. A general form of the Liouville operator for a dynamical system of the form (2.1) without random noise is

$$L_L(\vec{X})p = -\operatorname{div}_{\vec{X}}\left(\vec{F}(\vec{X})p\right). \tag{2.51}$$

For all deterministic systems the first term in the right-hand side of (2.49) can be reduced explicitly to

$$\int_{\mathbb{R}^N} \ln\left(1 + \frac{\delta p}{p_{\text{eq}}}\right) L_L \delta p = -\frac{1}{2}\int_{\mathbb{R}^N} \frac{\delta p^2}{p_{\text{eq}}^2} L_L \delta p + \mathcal{O}\left(\|\delta p(\vec{x}, t)\|^4\right). \tag{2.52}$$

The relation in (2.52) can be proven in the following way: taking into account (2.51), and the fact that $L_L p_{\text{eq}} = 0$, one writes

$$\begin{aligned}
\int_{\mathbb{R}^N} \ln\left(1 + \frac{\delta p}{p_{\text{eq}}}\right) L_L \delta p &= \int_{\mathbb{R}^N} \delta p \vec{F} \cdot \nabla \ln\left(1 + \frac{\delta p}{p_{\text{eq}}}\right) d\vec{x} \\
&= \frac{1}{2}\int_{\mathbb{R}^N} \frac{\vec{F}\cdot\nabla(\delta p^2)}{p_{\text{eq}} + \delta p} + \int_{\mathbb{R}^N} \frac{\delta p^2 \nabla\cdot\vec{F}}{p_{\text{eq}} + \delta p} \\
&= -\frac{1}{2}\int_{\mathbb{R}^N} \delta p^2 \nabla\cdot\left(\frac{\vec{F}}{p_{\text{eq}} + \delta p}\right) + \int_{\mathbb{R}^N} \frac{\delta p^2 \nabla\cdot\vec{F}}{p_{\text{eq}} + \delta p} \\
&= -\frac{1}{2}\int_{\mathbb{R}^N} \frac{\delta p^2 \nabla\cdot\vec{F}}{p_{\text{eq}} + \delta p} + \frac{1}{2}\int_{\mathbb{R}^N} \frac{\delta p^2 \nabla\cdot(\vec{F}\delta p)}{(p_{\text{eq}} + \delta p)^2} \\
&\quad + \int_{\mathbb{R}^N} \frac{\delta p^2 \nabla\cdot\vec{F}}{p_{\text{eq}} + \delta p} = \frac{1}{2}\int_{\mathbb{R}^N} \frac{\delta p^2 \vec{F}\cdot\nabla\delta p}{(p_{\text{eq}} + \delta p)^2} \\
&\quad + \frac{1}{2}\int_{\mathbb{R}^N} \frac{\delta p^2 \nabla\cdot\vec{F}}{p_{\text{eq}} + \delta p} - \frac{1}{2}\int_{\mathbb{R}^N} \frac{\delta p^2 p_{\text{eq}} \nabla\cdot\vec{F}}{(p_{\text{eq}} + \delta p)^2} \\
&= -\frac{1}{2}\int_{\mathbb{R}^N} \frac{\delta p^2}{p_{\text{eq}}^2} L_L \delta p + \mathcal{O}(\|\delta p\|^4).
\end{aligned} \tag{2.53}$$

Thus, from (2.52), for deterministic dynamical systems the first term on the right-hand side of (2.49) is $O(\|\delta p\|^3)$. Therefore, the first term in (2.50) is identically zero for a Liouville operator $L_L(\vec{X})$ in (2.51), and the evolution of the leading order entropy term in (2.50) becomes

$$\frac{d\tilde{P}}{dt} = \int_{\mathbb{R}^N} \frac{\delta p(\vec{X}, t)}{p_{\text{eq}}(\vec{X})} \delta L_{\text{ext}}(\vec{X}, t) p_{\text{eq}}(\vec{X})\, d\vec{X} = \delta E\left[\frac{\vec{L}_{\text{ext}}(\vec{X}) p_{\text{eq}}(\vec{X})}{p_{\text{eq}}(\vec{X})}\right] \cdot \delta\vec{F}(t), \tag{2.54}$$

and can be evaluated directly using the fluctuation-dissipation framework of Section 2.2 as stated in Theorem 2.1. While we have not done so here, it is not difficult to generalize this calculation to include the effect of noise in (2.1).

2.4. The fluctuation-dissipation theorem and ensemble predictions near equilibrium

As discussed in Chapter 1, ensemble predictions are the natural way to make predictions for complex chaotic nonlinear systems with many degrees of freedom. In such an ensemble prediction, the deterministic initial mean state $\vec{X}_0$ is replaced by an ensemble of M perturbed initial conditions, i.e. the initial data $\vec{X}_0$ is replaced by the perturbed initial data, $\vec{X}_0 + \delta\vec{X}_0^i$, $1 \leq i \leq M$. In general, it is very difficult to develop a theory for the outcome of such an individual ensemble prediction in detail. However, suppose we let the initial data $\vec{X}_0$ vary over the smooth equilibrium density, $p_{\text{eq}}(\vec{X})$, and ask the simpler question regarding the behavior of the expected value of the prediction ensemble over all initial data $\vec{X}$ sampled according to p_{eq} and perturbed by $\delta\vec{X}^i$; in the jargon of ensemble prediction, this is the statistical behavior of the "superensemble". Here we show that the fluctuation-dissipation Theorem 2.1 from Section 2.2 can be utilized to determine the mean statistical behavior of the ensemble prediction provided that the strength of the perturbations, $\delta\vec{X}^i$, is small enough.

To rephrase the above problem in mathematical terms, for each independent perturbation kick, $\delta\vec{X}^i$, of the initial data for (2.1), we need to compare the smooth density, $p_{\text{eq}}(\vec{X})$, with $p^\delta(\vec{X}, t)$, which satisfies the Fokker–Planck equation in (2.2)

$$\begin{aligned} \frac{\partial}{\partial t} p_t(\vec{X}) &= -\operatorname{div}_{\vec{X}}(\vec{F} p_t) + \sum_{ij} \frac{\partial^2}{\partial X_i \partial X_j}\left(\frac{Q_{ij}}{2} p_t\right), \\ p_t(\vec{X})\big|_{t=0} &= p_{\text{eq}}(\vec{X} + \delta\vec{X}_0^i). \end{aligned} \tag{2.55}$$

Note that in the present situation, the Fokker–Planck operator is unperturbed but the initial probability density is perturbed. To understand the average statistical effects of the ensemble prediction, we write $p^\delta(\vec{X}, t) = p_{\text{eq}} + \delta p$ and, as in (2.20), we need to calculate

$$\delta E[A(\vec{X})](t) = \int_{\mathbb{R}^N} A(\vec{X}) \delta p(\vec{X}, t)\, d\vec{X}. \tag{2.56}$$

Theorem 2.1 from Section 2.2 allows us to compute the response in (2.56) in general provided that $\delta \ll 1$ for a given time t. To achieve this, we utilize $\vec{L}_{\text{ext}} = -\frac{\partial}{\partial X_i}$, $1 \leq i \leq N$ and the specific choice $\delta\vec{F}(t) = \delta\vec{X}_0^i \delta(t)$ where $\delta(t)$ is the delta function in time. It is easy for the reader to check that this external perturbation corresponds formally to the perturbed initial value problem in (2.55) for the Fokker–Planck equation. With the above formal identification, the FDT in Section 2.2 applies and guarantees for $\delta \ll 1$,

$$\delta E[A(\vec{X})](t) = \vec{R}_A(t) \cdot \delta\vec{X}_0^i \tag{2.57}$$

with the response vector operator, $\vec{R}_A(t)$, given explicitly by

$$\vec{R}_A(\tau) = -\big\langle A\big(\vec{X}(t+\tau)\big) \nabla \ln p_{\text{eq}}\big(\vec{X}(t)\big)\big\rangle. \tag{2.58}$$

Notice that the formula in (2.57) is linear in the independent initial kicks, $\delta\vec{X}_0^i$, so that the formula in (2.57) generalizes immediately to

$$\delta E[A(\vec{X})](t) = \vec{R}_A(t)\cdot\delta\bar{\vec{X}}_0 \tag{2.59}$$

with $\delta\bar{\vec{X}}_0$ being the initial ensemble perturbation mean, i.e.

$$\delta\bar{\vec{X}}_0 = \frac{1}{M}\sum_{i=1}^{M}\delta\vec{X}_0^i. \tag{2.60}$$

The above formulas apply to perturbed ensemble prediction for the general equation in (2.1).

In order to gain some practical insight, we assume for simplicity that $\vec{F}(\vec{X})$ has the simplified structure in (2.29)–(2.30) with the Gaussian invariant measure in (2.31). In Chapter 1, we have mentioned that in ensemble predictions, it is reasonable to monitor a finite number of moments in the ensemble prediction; for example, the first four moments, $\langle X_i^r(t)\rangle$, $1\le r\le 4$, of individual components contain a reasonable amount of practical information to develop non-Gaussian estimators for the ensemble prediction [**1, 11, 14**]. Thus, with (2.34) in the present setting and (2.57)–(2.60) we have the general formulas for the perturbed moments in the ensemble prediction,

$$\delta E[X_i^r](\tau) = \theta\langle X_i^r(t+\tau)\vec{X}(t)\rangle\cdot\delta\bar{\vec{X}}_0 \tag{2.61}$$

for any $1\le i\le N$ and any r. Note that the estimate for the "superensemble" averaged rth moment involves higher moment correlations at different times. This result is even interesting at time $t=0$ where the averaged odd perturbed moments are nonzero, for example

$$\begin{aligned}\delta E[X_i](0) &= \theta\langle X_i(0)\vec{X}(0)\rangle\cdot\delta\vec{X}_0 = (\delta\vec{X}_0)_i,\\ \delta E[X_i^3](0) &= \theta\langle X_i^3(0)\vec{X}(0)\rangle\cdot\delta\vec{X}_0 = \theta\langle X_i^4(0)\rangle(\delta\vec{X}_0)_i\\ &= 3\theta\langle X_i^2(0)\rangle^2(\delta\vec{X}_0)_i = \frac{3}{\theta}(\delta\vec{X}_0)_i.\end{aligned} \tag{2.62}$$

On the other hand, all even moments vanish at $t=0$ since

$$\delta E[X_i^{2k}](0) = \theta\langle X_i^{2k}(0)\vec{X}(0)\rangle\cdot\delta\vec{X}_0 = 0. \tag{2.63}$$

Thus, the perturbed superensemble distribution near equilibrium starts out with a nontrivial ensemble mean, and nonzero third moment, the skewness. Which of these factors is the controlling feature of the average ensemble prediction near equilibrium depends crucially on the decay of the higher correlations in (2.61) in time compared with the time decay of the lower order correlations. These decay times reflect the underlying nonlinear statistical features of the dynamical system in (2.29) in statistical equilibrium and can be monitored directly for the unperturbed climate without doing many families of ensemble predictions directly! This is a powerful potential use of the nonlinear features of FDT in understanding ensemble predictions. This natural use of nonlinear functionals in the FDT also motivated our discussion above.

2.5. Computational strategies for applying and testing the validity of the fluctuation-dissipation theorem

One of the attractive features of the fluctuation-dissipation theorem described in detail in Section 2.2 is that the predictions of this theorem for the perturbed mean response, $\delta E[A(\vec{X})]$, involve only computations of the linear response operator, $\vec{R}_A(\tau)$, defined by the formula for nonlinear correlations,

$$\vec{R}_A(\tau) = \left\langle A\Big(\vec{X}(t+\tau)\frac{\vec{L}_{\text{ext}}p_{\text{eq}}}{p_{\text{eq}}}(\vec{X}(t))\Big)\right\rangle. \tag{2.64}$$

The important feature of the correlation function in (2.64) is that, according to (2.24), it only needs to be computed for the unperturbed climate. If the dynamical system in (2.1) is ergodic and mixing with exponential decay of correlations, the correlation function $\vec{R}_A(\tau)$ can be calculated through the time average of a typical solution

$$\vec{R}_A(\tau) = \lim_{T\to\infty}\frac{1}{T}\int_{T_0}^{T_0+T} A\big(\vec{X}(t+\tau)\big)\frac{\vec{L}_{\text{ext}}p_{\text{eq}}}{p_{\text{eq}}}\big(\vec{X}(t)\big)\,dt. \tag{2.65}$$

All three of the dynamical system models in (2.32), the TBH, K-Z and IL96 models, are strongly mixing and ergodic according to the numerical test of decay of correlations; see Section 2.6 below for numerical confirmation with 40 active modes in each of these models. Similar statements apply to the equations for truncated quasi-geostrophic flow with topography [**17**]. A detailed discussion of all these results and examples can be found in [**18**]. With the tacit assumption in (2.65), the response operator $\vec{R}_A(\tau)$ can be calculated by integrating a typical individual trajectory of (2.1) for a large time $T \gg 1$ and using the right hand side of (2.65) to calculate an approximation to $\vec{R}(\tau)$; the time T_0 is the time in which it takes the trajectory with a given initial data to adjust to the chaotic attractor where numerical averaging begins. Examples of such a procedure are presented below in Sections 2.6 and 2.8 as well as in many of the references for this chapter. For the examples in Sections 2.6 and 2.8, the initial time T_0 is chosen to be 10000 time units while the averaging window T is 100000 time units.

How can we test the validity of the predictions of the fluctuation-dissipation theorem for a specific nonlinear dynamical system from (2.1) or with the special form in (2.29), (2.30)? Recall that FDT in Section 2.2 states that for appropriate external perturbations, defined through $\delta\vec{F}(t)$ and $\vec{L}_{\text{ext}}$,

$$\delta E[A(\vec{X})](t) = \int_0^t \vec{R}_A(t-t')\cdot\delta\vec{F}(t')\,dt' \equiv \mathcal{R}_{\text{FDT}}(t)\delta\vec{F}. \tag{2.66}$$

For the special case of time independent forcing $\delta\vec{F}$ we have

$$\mathcal{R}_{\text{FDT}}(t) = \int_0^t \vec{R}_A(t')\,dt'. \tag{2.67}$$

Intuitively, to test the validity of the predictions of FDT, we need to supply specific external perturbations, defined by $\delta\vec{F}(t)$ and the vector-field $\vec{a}(\vec{X})$ from (2.13). We integrate the nonlinear dynamics in (2.1) explicitly with $\delta \ll 1$ up to time t and attempt to find the ideal response operator, $\mathcal{R}_{\text{Ideal}}(t)$ which is the best linear approximation to the perturbed dynamics for $\delta \ll 1$ for the time t. Then one can compare $\mathcal{R}_{\text{Ideal}}(t)$ and $\mathcal{R}_{\text{FDT}}(t)$ directly to see which features of $\mathcal{R}_{\text{Ideal}}(t)$ are captured by $\mathcal{R}_{\mathrm{FDT}}(t)$. As described at the beginning of this section, $\mathcal{R}_{\mathrm{FDT}}(t)$ is readily calculated in a comparatively inexpensive fashion; on the other hand, the determination of the ideal response operator, $\mathcal{R}_{\mathrm{Ideal}}(t)$, requires much more extensive numerical calculations and some innovative algorithms. This is the price which must be paid in developing a computational laboratory to test the validity of FDT compared with the ideal response. Next, we summarize a computational strategy for calculating the ideal response operator for any dynamical system from (2.1) for the case of steady forcing where $\delta\vec{F}$ is time-independent. The algorithm is essentially due to Gritsoun and Dymnikov **[9]**; we apply it subsequently in Sections 2.6 and 2.8 below. The algorithm will be less computationally intensive for the infinite time ideal response, $\mathcal{R}_{\mathrm{Ideal}}(\infty)$, where time averaging of an individual solution is utilized as compared with the finite time ideal response operator, $\mathcal{R}_{\mathrm{Ideal}}(t)$, where as ensemble of solutions on the attractor is needed.

2.5.1. A numerical algorithm for the ideal response operator. Below we present a step-by-step description of the algorithm for constructing the ideal response operator.

- First we make the basic linearity assumption: for a given functional A, and small time-independent forcing $\delta\vec{F}$, the difference between the unperturbed expectation $\langle A\rangle$ and the finite-time perturbed expectation $\langle A\rangle(\delta\vec{F},t)$ is given by

$$\langle A\rangle(\delta\vec{F},t)-\langle A\rangle=\langle\delta A\rangle(\delta\vec{F},t)=\mathcal{R}_{\mathrm{Ideal}}(t)\delta\vec{F}, \tag{2.68}$$

which is also the definition for the finite-time response operator $\mathcal{R}_{\mathrm{Ideal}}(t)$. The angled brackets in (2.68) denote the ensemble average. For the infinite time ideal response operator, the formula in (2.68) becomes

$$\bar{A}(\delta\vec{F})-\bar{A}=\bar{\delta A}(\delta\vec{F})=\mathcal{R}_{\mathrm{Ideal}}(\infty)\delta\vec{F}, \tag{2.69}$$

where the overbar denotes the average of a long time series of a single mixing solution as discussed earlier in this section.

- The next step is to compute the equilibrium statistical state of the unperturbed dynamical system. For the time dependent linear response operator $\mathcal{R}_{\mathrm{Ideal}}(t)$ we generate an initial ensemble of solutions, usually with 10000 members, and let it propagate freely until it reaches statistical equilibrium. By averaging over the ensemble, we calculate the mean state of a functional $\langle A\rangle$. For the infinite time linear response operator $\mathcal{R}_{\mathrm{Ideal}}(\infty)$ we compute long time series of a single solution, and then evaluate $\bar{A}$ by the time average.
- Then the dynamical system is perturbed by external forcing of the form

$$\delta\vec{F}(k,r)=r\vec{e}_k, \tag{2.70}$$

for different values of r and k, where $\vec{e}_k$ is the kth orthonormal basis vector in the phase space, and r is a small positive scalar parameter which characterizes the strength of the small forcing. For each pair of r and k, the time-dependent response is obtained by forcing the ensemble out of equilibrium and recording the response of $\langle A\rangle(k,r,t)$; the infinite time response is obtained by re-running the dynamical system with external forcing and recording the deviation of the mean $\bar{A}(k,r)$.
- For a given k, the response is linear for a certain range of $r\in[-a,a]$, if the corresponding vectors of time-dependent response $\langle\delta A\rangle(k,r,t)$ or infinite time response $\bar{A}(k,r)$ for $\delta\vec{F}(k,r)$, $-a\le r\le a$ are roughly collinear. The threshold amplitude a is chosen experimentally and depends on the nature of the dynamical

model being tested. Thus, for a given k, the kth column of $\mathcal{R}_{\text{Ideal}}(t)$ or $\mathcal{R}_{\text{Ideal}}(\infty)$ is determined as the least squares approximation over the number of forcing—linear response outcomes with different r's belonging to $[-a, a]$. This is done by the minimization of

$$S_k(t) = \sum_i |\langle \delta A \rangle (k, r_i, t) - r_i R_k(t)|^2 \tag{2.71}$$

over $R_k(t)$, which is the kth column of $\mathcal{R}_{\text{Ideal}}(T)$ for the time-dependent linear response, and

$$S_k(\infty) = \sum_i |\delta \bar{A}(k, r_i) - r_i R_k(\infty)|^2 \tag{2.72}$$

for the infinite time response, where the summation is done over the discrete set $\{r\}_i \subset [-a, a]$.

- The full linear response operators $\mathcal{R}_{\text{Ideal}}(t)$ and $\mathcal{R}_{\text{Ideal}}(\infty)$ are then constructed column-by-column through iterating the above steps over all independent directions of the phase space $\vec{e}_k$. The errors $K(r_i, t)$ and $K(r_i, \infty)$ are then determined by their formulas

$$K(k, r_i, t) = \frac{|\langle \delta \vec{A} \rangle (k, r_i, t) - R_k(t) r_i|}{|R_k(t) r_i|}, \tag{2.73}$$

$$K(k, r_i, \infty) = \frac{|\delta \bar{\vec{A}}(k, r_i) - R_k(\infty) r_i|}{|R_k(\infty) r_i|}. \tag{2.74}$$

The errors $K(k, r_i, t)$ and $K(k, r_i, \infty)$ actually measure the extent in which the dynamic response remains nonlinear for the given threshold of forcing defined by r_i.

2.6. Three chaotic nonlinear systems demonstrating the validity of FDT

Here we study the validity of FDT for the three different strongly mixing chaotic dynamical systems discussed earlier in (2.32) which have the structure in (2.29)–(2.31). We study each of these models with forty degrees of freedom. These three models are the following:

- The TBH model

$$\begin{gathered} (u_\Lambda)_t = -\tfrac{1}{2} P_\Lambda (u_\Lambda^2)_x, \\ P_\Lambda f = \sum_{|k| \le \Lambda} \hat{f}_k e^{ikx}, \quad \hat{f}_{-k} = \bar{\hat{f}}_k, \quad \Lambda = \frac{J}{2}. \end{gathered} \tag{2.75}$$

- The K-Z model

$$\begin{gathered} \frac{du_j}{dt} = -\frac{1}{\Delta x}(F_{j+1/2} - F_{j-1/2}), \quad j = 0, 1, \ldots, J \\ \text{with } F_{j+1/2} = \tfrac{1}{6}(u_j^2 + u_j u_{j+1} + u_{j+1}^2) \end{gathered} \tag{2.76}$$

  and periodic boundary conditions $u_J = u_0$.

- The IL96 model

$$\frac{du_j}{dt} = (u_{j+1} - u_{j-2}) u_{j-1}, \quad j = 0, 1, \ldots, J \tag{2.77}$$

  with periodic boundary conditions $u_J = u_0$.

As discussed in Section 2.2, all three models conserve energy and have a family of Gaussian invariant measures as given in (2.31). For these three dynamical systems, the Gaussian invariant measure, $p_{\text{eq}}(\vec{X})$, leads to following predictions for the climatology:

(2.78)
- Zero mean $\langle \hat{u}_k \rangle = 0$,
- Equipartition of energy $\langle |\hat{u}_k|^2 \rangle = \text{const}, \ 0 \le |k| \le \frac{J}{2}$

where $\hat{u}_k$ are the Fourier coefficients. In (2.78), as in earlier sections, the average is defined by

$$\langle f(\vec{X}) \rangle = \int_{\mathbb{R}^N} f(\vec{X}) p_{\text{eq}}(\vec{X})\, dX. \tag{2.79}$$

All three of the dynamical systems are ergodic and strongly mixing with exponential decay of correlations provided the number of degrees of freedom is large enough; the value of $J = 40$ degrees of freedom utilized here is more than sufficient according to the studies in [**16**] for (2.75), (2.76) and [**1**] for (2.77). Thus, as discussed at the beginning of Section 2.5, the ensemble average, $\langle f \rangle$, in (2.79) can be approximated with high accuracy by the time average over an individual trajectory,

$$\overline{f(\vec{X})} = \frac{1}{T} \int_{T_0}^{T_0+T} f(\vec{X}(t))\, dt \tag{2.80}$$

where $T \gg 1$. Below we simply equate $\overline{f}$ with $\langle f \rangle$ for both the calculation of the climatology and the correlation functions from (2.64) and (2.65) needed in FDT. We verify below that all three models have a "climatology" where the predictions of $p_{\text{eq}}(\vec{X})$ in (2.78) are satisfied with excellent accuracy. Thus, superficially, all three models have the same climatology.

However, our interest here is the response predicted by FDT and its validity compared with the ideal response operator discussed at the end of Section 2.5. Recall from the earlier discussion around (2.33)—(2.35) that for constant external forcing δF added to any of the three systems in (2.75)—(2.77) and for the special linear functionals, $\vec{A}(\vec{X}) = \vec{X}$, the FDT predicts the response operator,

$$\mathcal{R}_{\text{FDT}}(t) = \theta \int_0^t \langle \vec{X}(\tau + t_0) \otimes \vec{X}(t_0) \rangle\, d\tau. \tag{2.81}$$

While the above three dynamical systems have the same climatology statistically, they differ widely in their predicted FDT response operator with the Fourier space correlation functions having the following features verified below:

(2.82)
- The TBH model has the largest correlation times at the largest scales or equivalently smallest wave numbers with very little oscillation in the individual correlation functions.
- The K-Z model has much more oscillatory decaying correlations with a competition of large correlation times between large and small scale modes changing with the parameter t for $\mathcal{R}_{\text{FDT}}(t)$.
- The IL96 model has essentially identical non-oscillatory correlation functions independent of wave numbers.

With the qualitative behavior in (2.82), the three models have radically different response operators $\mathcal{R}_{\text{FDT}}(t)$ as defined in (2.81). Thus, these models are an attractive laboratory for testing the validity of FDT. First, with A), the TBH model is expected to have the most significant linear response on the largest scales or the smallest wave number Fourier modes essentially independent of time. With B) for the K-Z model, as the time t changes, the response operator, $\mathcal{R}_{\text{FDT}}(t)$, is dominated by a combination of behavior reflecting both the largest and smallest scales in the system with bifurcations occurring in this behavior in time. In contrast to both the behavior for TBH and K-Z, the IL96 model behaves essentially like a gas of identical particles with essentially the same response, $\mathcal{R}_{\text{FDT}}(t)$, on all scales for a given time. Thus, these three examples provide a rich laboratory for testing the validity of FDT in an idealized setting with diverse phenomena predicted by FDT in a wide variety of circumstances.

Below, in contrast to all the earlier studies, we will calculate the entire ideal response operator, $\mathcal{R}_{\text{Ideal}}(t)$, by the procedure discussed in Section 2.5 and compare it directly in a variety of ways with $\mathcal{R}_{\text{FDT}}(t)$ from (2.81), also calculated from time averaging as described in Section 2.5 above. For the three models used in the present study, the ensemble size for the construction of $\mathcal{R}_{\text{Ideal}}(t)$ has 10000 members while the time averaging window, T, is 100000 units after an initial adjustment time, T_0, with $T_0 = 10000$. The time integration of all three dynamical systems is done by the Adams–Bashforth multistep method with a time step of $\Delta T = 1/64$ units. The initial conditions for each model are generated at random from the uniform distribution on a constant energy surface. The values of energy are: $E = 0.1$ for TBH model, $E = 100$ from K-Z model, and $E = 10$ for IL96 model. These constant energy values are chosen so that the decorrelation time scales for largest Fourier modes are of the same order in magnitude for all three models, which also justifies the choice of fixed integration time step and averaging window.

All three of the models in (2.75)—(2.77) are translation invariant so that in physical space both $\mathcal{R}_{\text{FDT}}(t)$ and $\mathcal{R}_{\text{Ideal}}(t)$ can be written as discrete convolution operators

$$\begin{aligned}\mathcal{R}_{\text{FDT}}(t)\delta\vec{F} &= \sum_{j=0}^{J} \mathcal{R}_{\text{FDT},i-j}(t)\delta F_j,\\ \mathcal{R}_{\text{Ideal}}(t)\delta\vec{F} &= \sum_{j=0}^{J} \mathcal{R}_{\text{Ideal},i-j}(t)\delta F_j.\end{aligned} \tag{2.83}$$

Below we compare linear response operators through their kernels, $\mathcal{R}_{\text{FDT},i}(t) = g$ and $\mathcal{R}_{\text{Ideal},i}(t) = h$ by measuring their L^2 relative error

$$2\frac{\|g-h\|}{\|g\|+\|h\|} \tag{2.84}$$

and their correlation function

$$C(g,h) = \frac{(g,h)}{\|g\|\|h\|}, \tag{2.85}$$

where $(g,h) = \sum_{j=0}^{J} g_j h_j$ and $\|g\| = (g,g)^{1/2}$. Note by the Cauchy-Schwarz inequality that the correlation function satisfies $-1 \le C(g,h) \le 1$ and that $C(g,h) = 1$ means that $g = \alpha h$ with $\alpha > 0$, i.e. g and h are collinear. Thus, a value of $C(g,h)$

near 1 quantifies the extent that g and h are nearly collinear irrespective of their amplitudes.

With (2.83), it follows that in Fourier space, the linear response operator automatically splits into an orthogonal sum of real two-dimensional operators defined for each Fourier wave number, k, with $0 \leq |k| \leq J/2$; with this observation, it is easy to calculate all the singular values of the response operator and their associated singular vectors. Below, the leading few largest response eigenvalues for $\mathcal{R}_{\text{Ideal}}(t)$ and $\mathcal{R}_{\text{FDT}}(t)$ are recorded as well as the wavenumbers of the associated singular vectors. Similar numerical consideration and quantitative measures as described in the preceding paragraphs are also utilized in discussing the damped forced Lorenz 96 model in Section 2.8 below and will not be repeated here. With all of the above background, we are finally ready to discuss the climatology and validity of FDT for the three models in (2.75), (2.76) and (2.77).

2.6.1. The validity of FDT for the TBH model. First, we report the climatology for TBH. The variance spectrum of the TBH model is shown in Figure 2.1. The variance spectrum is defined as the set of time averaged variances for the spectral coefficients

$$\text{Var}(\hat{u}_k) = \overline{|\hat{u}_k(t) - \bar{\hat{u}}_k|^2}, \tag{2.86}$$

where the overbar denotes the time average, such that $\bar{\hat{u}}_k$ is the time average of $\hat{u}_k(t)$. The variance spectrum in Figure 2.1 is uniform as predicted by (2.78).

The time autocorrelation functions are shown in Figure 2.2 for the wavenumbers $k = 1, \ldots, 4$. A general formula for the autocorrelation function of a variable $a(t)$ is given by

$$\text{Corr}(\tau) = \frac{\overline{(a(t) - \bar{a})(a(t+\tau) - \bar{a})}}{\overline{(a(t) - \bar{a})^2}}, \tag{2.87}$$

where $\bar{a}$ is the time average of $a(t)$. In practice the time averaging window T is finite but large (10^5 time units in our simulations). The autocorrelation function in (2.87) measures mixing in a dynamical system, and in some sense can be regarded as an indicator of how fast a variable "forgets" about its behavior in the past.

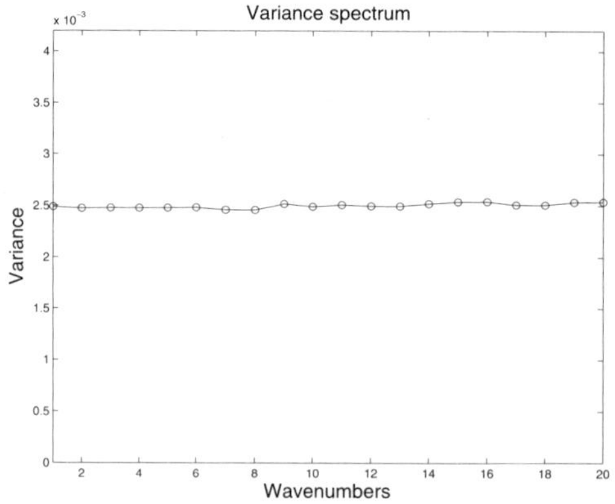

FIGURE 2.1. The variance spectrum for the Burgers–Hopf model.

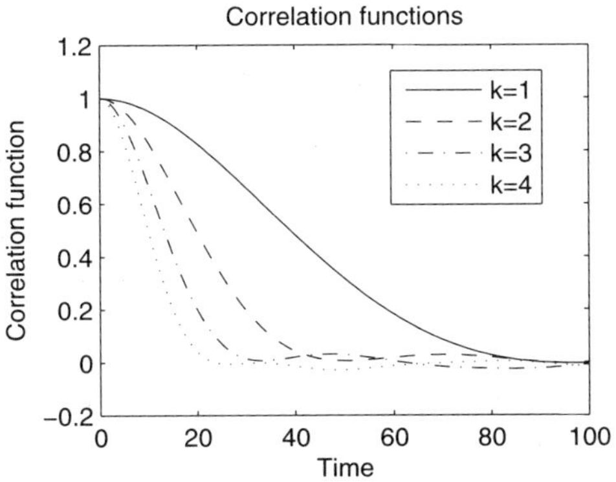

FIGURE 2.2. The time autocorrelation functions for the wavenumbers $k = 1, \ldots, 4$, Burgers–Hopf model. Solid line—$k = 1$, dashed line—$k = 2$, dot-dashed line—$k = 3$, dotted line—$k = 4$.

A distinct scaling for correlation function behavior can be observed as mentioned in (2.82)—that is, the time autocorrelation functions evolve slowly for large scale wavenumbers, whereas small scale wavenumbers mix faster. This is a general feature of TBH which is predicted by simple scaling arguments [**15, 16**].

The correlation (or, rather, decorrelation) time is defined as a typical time of the decay of the autocorrelation function. Usually it is given through the time integral of the autocorrelation function. Here we look at two different measures of the correlation time: one is the integral of the autocorrelation function

$$T = \int_0^\infty \operatorname{Corr}(\tau)\, d\tau, \tag{2.88}$$

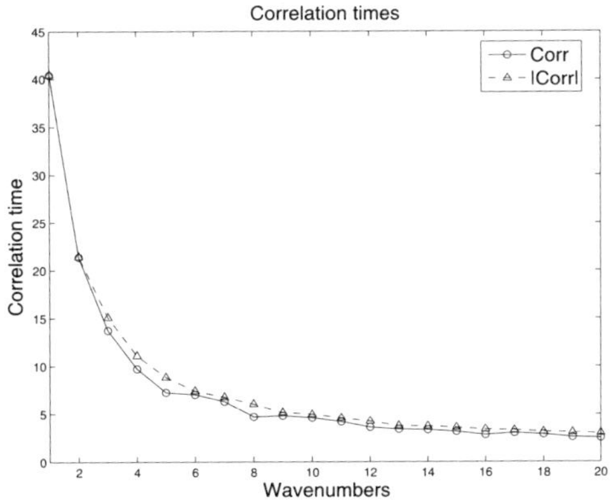

FIGURE 2.3. The spectra of correlation times (integrals of correlation functions) for the Burgers–Hopf model. Solid line with circles: the integrals are taken with the time correlation functions directly. Dashed line with triangles: the integrals are taken with the absolute values of the correlation functions.

and the other is the integral of the absolute value of the autocorrelation function

$$T_{\text{abs}} = \int_0^\infty |\operatorname{Corr}(\tau)|\, d\tau. \tag{2.89}$$

Clearly $T_{\text{abs}} \geq T$ and T_{abs} accounts for all the oscillations in the correlation function. The two types of correlation time spectra are presented in Figure 2.3: the correlation times shown by solid lines in Figure 2.3 are obtained by integrating the time autocorrelation functions, whereas the correlation times shown by dashed lines in Figure 2.3 are obtained by integrating the absolute value of the corresponding correlation function. We also see a decay of correlation times as wavenumbers become higher. Note that the T and T_{abs} do not differ significantly at all scales. This means that negative portions of autocorrelation functions at all scales are close to zero as noted in (2.82).

Now, we return to the validity of FDT for the TBH model. The ideal linear response operator is constructed for the simplest functional $\vec{A}(\vec{x}) = \vec{x}$. The FDT linear response operator is constructed in Fourier space.

2.6.1.1. *Time evolution of singular values.* First, we begin with a remark which also applies to the K-Z model. In the weakly perturbed model, the energy is

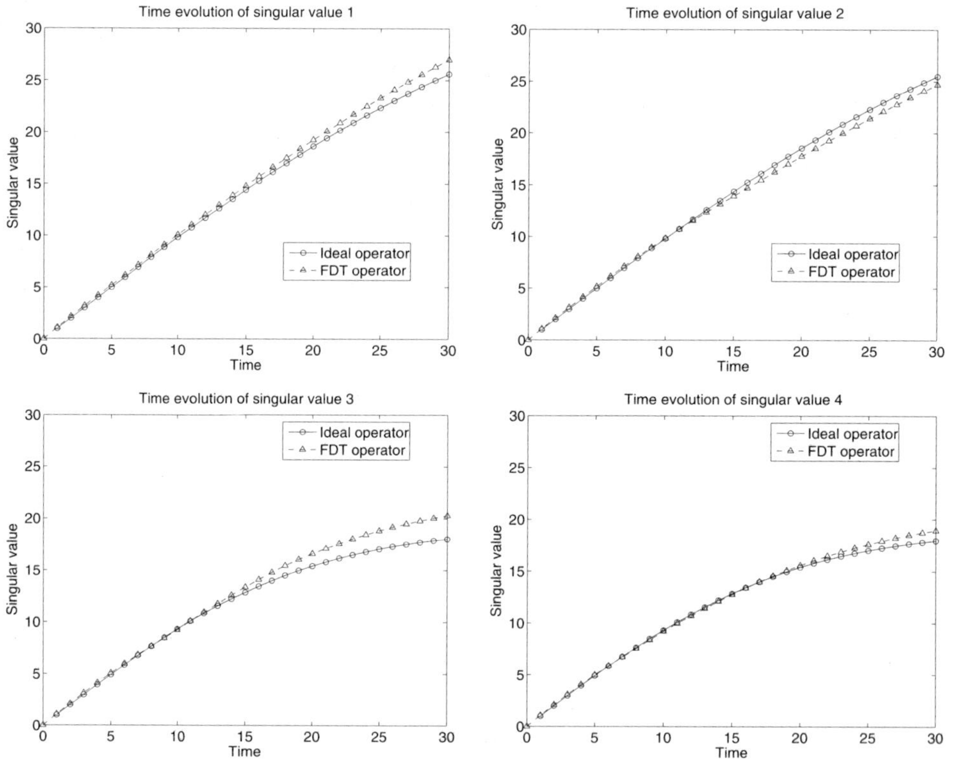

FIGURE 2.4. The time evolution of singular values 1, 2, 3 and 4 for the Burgers–Hopf model. Solid line—singular values of the ideal linear response operator. Dashed line—the singular values of the FDT response operators.

constantly pumped into an otherwise conservative dynamical system. Because of that, the largest singular value of the ideal response operator grows linearly with time and falls onto the zero Fourier mode. Remember that the zero Fourier mode is a conserved quantity in the unperturbed TBH system and is set to zero by initial conditions. Due to this fact, it is apparent that the FDT framework can not possibly capture the evolution of the largest singular value for trivial reasons. On the other hand, it can be easily seen that if a perturbation does not fall onto any nonzero Fourier mode, there is no response in the zero Fourier mode. Thus, the ideal operator is constructed without perturbing the zero Fourier mode, which is completely compatible with the FDT.

The time evolution of the largest singular values 1–4 is shown in Figure 2.4. Each picture contains two curves: solid—the result for the ideal operator, and dashed—the result for the FDT operator. We can see that the time evolution of FDT singular values repeats the time evolution of the ideal response singular values with very minor discrepancies, for all displayed results. This is good evidence that the FDT framework is valid for the current model. We also observe a distinct difference in the magnitude of the response between different singular values, which is a direct consequence of correlation time scaling shown in Figures 2.2 and 2.3. Below we study the linear response in detail for the two time slices—$T = 10$ and $T = 20$.

2.6.1.2. *The response operators for time* $T = 10$. Here we show more extensive results for the time slice $T = 10$. In Figure 2.5 we display the ideal response operators and the FDT approximations in physical space with their main diagonals in the Fourier space. The general response structure in the Fourier space is concentrated on the main diagonal, but we do not show the imaginary part of the diagonal due to its similarity with the real part. This is common for both ideal and approximate FDT operators. We can see almost no visual differences between the ideal operator and the FDT operator, which means that the FDT framework is valid for the current model. The response in physical space is peaked at the

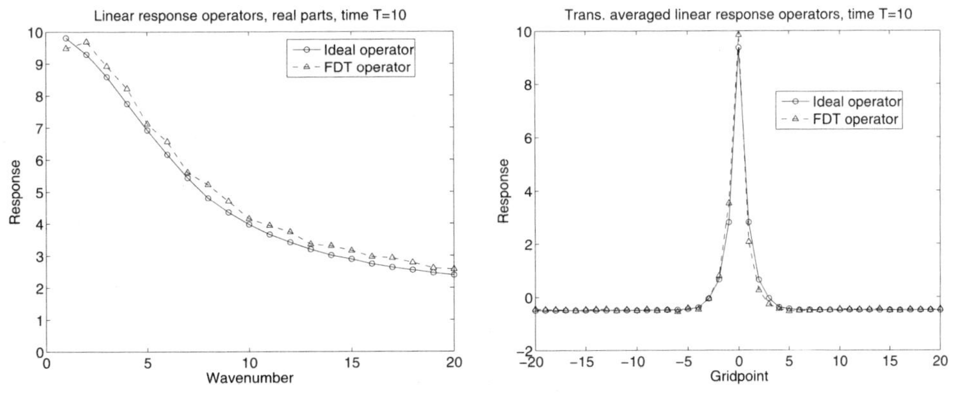

FIGURE 2.5. Left—real part of the main diagonal of linear response operator in Fourier space. Right—Translationally averaged linear response operators, in physical space. Solid line—ideal operator, dashed line—FDT operator. Time $T = 10$, the Burgers–Hopf model.

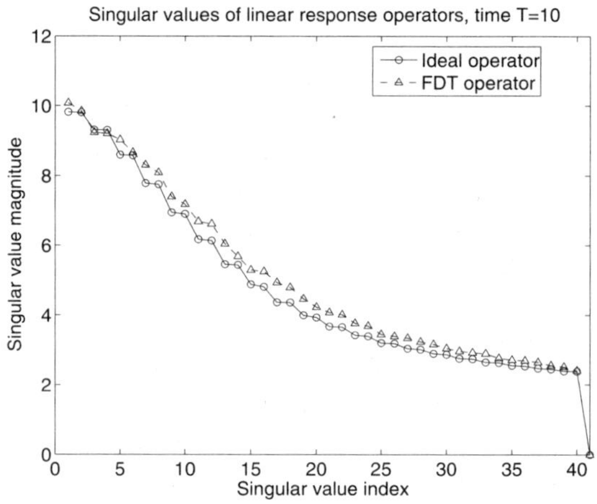

FIGURE 2.6. The spectrum of singular values for linear response operators. Solid line—ideal operator, dashed line—FDT operator. Time $T = 10$, the Burgers–Hopf model.

point where it is perturbed, however neighboring gridpoints also respond to the perturbation. The spectra of singular values for the time $T = 10$ are shown in Figure 2.6, which contains two curves: the solid curve denotes the results for the ideal linear response operator obtained by perturbing the system and measuring the response; the dashed curve shows the results for the FDT approximation of the response operator. As we can see, the FDT spectra of singular values capture the actual spectrum very well.

In Figure 2.7 we show the errors $K(k, r, T)$ from constructing the ideal response operator, $\mathcal{R}_{\text{Ideal}}(t)$. One can see that the error distribution across wavenumbers is highly nonuniform. The errors tend to concentrate on small scales, leaving the large scales essentially error-free. The reason for this behavior is the following: in this translation-invariant system the singular perturbations and singular responses are aligned with Fourier modes. This results in the situation where perturbations with large singular values produce adequate noise-free responses, thus inducing small errors, while responses from perturbations with small singular numbers are small enough to be non-distinguishable from the noise.

In Table 2.1 we show the errors (2.84) and correlations (2.85) for the time $T = 10$. The errors and correlations are presented for the real and imaginary parts of main diagonals of linear response operators in the Fourier space, for the translationally averaged linear response operators in physical space, and for the spectra of singular values. Since the errors are rather negligible for all measurements, the

TABLE 2.1. The errors and correlations for linear response operators at time $T = 10$ for the Kruskal–Zabusky model.

	L^2 error	$C(g, h)$
Fourier diag.	0.06501	0.9998
Phys. trans. av.	0.06332	0.9994

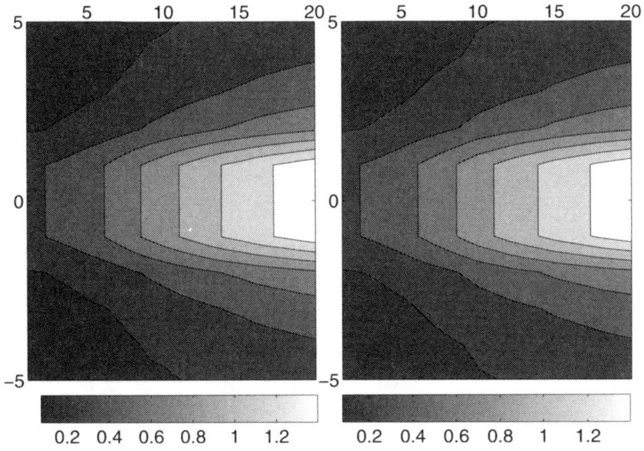

FIGURE 2.7. The errors $K(k, r, T)$ are shown as contour plots with k along horizontal axis and r as % of the energy along vertical axis. Horizontal axis denotes wavenumbers, real and imaginary perturbations are shown as separate right and left plots. Time $T = 10$, the Burgers–Hopf model.

ideal and FDT results are expected to be collinear, and Table 2.5 confirms this. In Table 2.2 we summarize the results for the time $T = 10$. These two tables correspond to the ideal and FDT results, respectively. Each table contains the following fields: SW—contains singular values, W_p—the wavenumber of corresponding singular perturbation, W_r—the wavenumber of corresponding singular response, and PD—the phase difference between the perturbation and the response. As we can see, all wavenumbers are captured perfectly by the FDT method. The singular vectors for both the ideal and FDT operators fall onto wavenumbers 1, 2, 3 and 4, which is a direct consequence of the correlation time scaling for TBH shown in

TABLE 2.2. The overview of results at time $T = 10$ for the Burgers–Hopf model, first 8 singular values. Left—ideal operator, right—FDT. Notations: SV—singular values, W_p—wavenumber on which singular perturbation is projected, W_r—wavenumber on which singular response is projected, PD—phase difference between singular perturbation and singular response.

N	SV	W_p	W_r	PD (deg.)
1	9.8176	1	1	0.0
2	9.8078	1	1	0.0
3	9.3089	2	2	0.0
4	9.3033	2	2	−0.0
5	8.5971	3	3	0.0
6	8.5767	3	3	0.0
7	7.7840	4	4	−0.0
8	7.7530	4	4	−0.0

N	SV	W_p	W_r	PD (deg.)
1	10.0792	1	1	−1.7
2	9.8469	2	2	3.0
3	9.2321	1	1	1.7
4	9.2040	2	2	−2.9
5	9.0244	3	3	4.6
6	8.6655	3	3	4.5
7	8.3065	4	4	−5.2
8	8.0910	4	4	−5.3

Figures 2.2 and 2.3. The phase difference here is captured rather well and it is essentially zero in all cases.

2.6.1.3. *The response operators for time $T = 20$.* Here we show more extensive results for the time slice $T = 20$. In Figure 2.8 we display the ideal response operators and the FDT approximations in physical space, with their main diagonals in the Fourier space. The general response structure in the Fourier space is concentrated on the main diagonal, but we do not show the imaginary part of the diagonal due to its similarity with the real part, like in the previous case with $T = 10$. This behavior is common for both ideal and approximate FDT operators. We can see almost no visual differences between the ideal operator and the FDT operator, which means that the FDT framework is valid for the current model. The spectrum of singular values for the time $T = 20$ is shown in Figure 2.9, which contains two curves: the solid curve denotes the results for the ideal linear response operator obtained by perturbing the system and measuring the response; the dashed curve shows the results for the FDT approximation of the response operator. As we can see, the FDT spectra of singular values capture the actual spectrum very well.

In Figure 2.10 we show the errors $K(k, r, T)$ for different perturbation-response methods of constructing ideal linear response operator. One can see that the error distribution across wavenumbers is highly nonuniform. As noted earlier for $T = 10$, the errors tend to concentrate on small scales, leaving large scales essentially error-free.

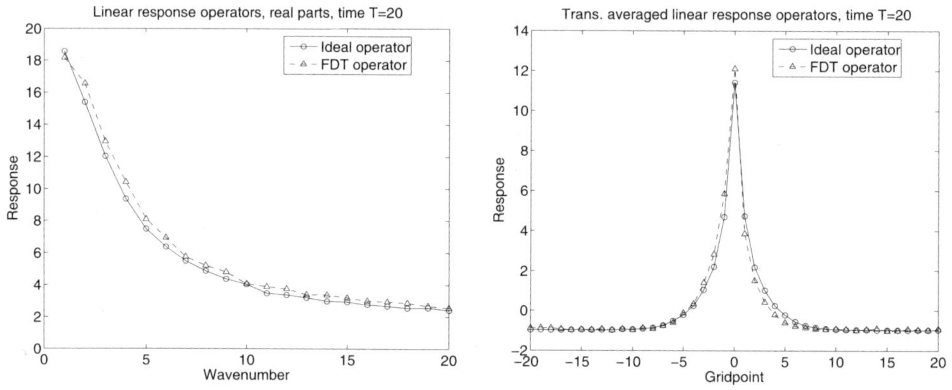

FIGURE 2.8. Left—real part of the main diagonal of linear response operator in Fourier space. Right—Translationally averaged linear response operators, in physical space. Solid line—ideal operator, dashed line—FDT operator. Time $T = 20$, the Burgers–Hopf model.

TABLE 2.3. The errors and correlations for linear response operators at time $T = 20$ for the Burgers–Hopf model.

	L^2 error	$C(g, h)$
Fourier diag.	0.06916	0.9988
Phys. trans. av.	0.141	0.9912

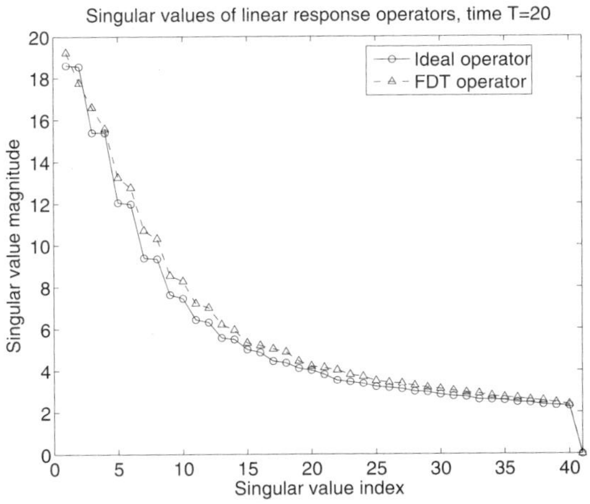

FIGURE 2.9. The spectrum of singular values for linear response operators. Solid line—ideal operator, dashed line—FDT operator. Time $T = 20$, the Burgers–Hopf model.

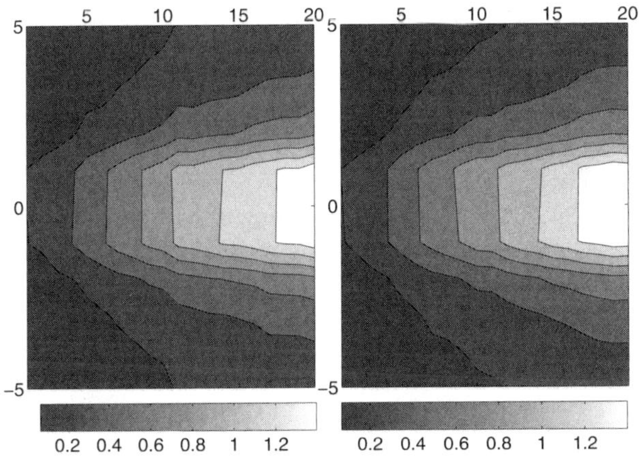

FIGURE 2.10. The errors $K(k, r, T)$ are shown as contour plots with k along horizontal axis and r as % of the energy along vertical axis. Horizontal axis denotes wavenumbers, real and imaginary perturbations are shown as separate right and left plots. Time $T = 20$, the Burgers–Hopf model.

In Table 2.3 we show the errors (2.84) and correlations (2.85) for the time $T = 20$. The errors and correlations are presented for the real and imaginary parts of main diagonals of linear response operators in the Fourier space, for the translationally averaged linear response operators in physical space, and for the spectra of singular values. Like for the time $T = 10$, the errors are rather negligible for all measurements, and the ideal and FDT results are collinear. In Table 2.4 we summarize both the ideal and FDT results for the time $T = 20$. As we can see, all wavenumbers are captured perfectly by the FDT method, which is surprising given

TABLE 2.4. The overview of results at time $T = 20$ for the Burgers–Hopf model, first 8 singular values. Left—ideal operator, right—FDT. Notations: SV—singular values, W_p—wavenumber on which singular perturbation is projected, W_r—wavenumber on which singular response is projected, PD—phase difference between singular perturbation and singular response.

N	SV	W_p	W_r	PD (deg.)
1	18.6077	1	1	−0.0
2	18.5443	1	1	0.0
3	15.3960	2	2	−0.1
4	15.3923	2	2	−0.1
5	12.0540	3	3	−0.2
6	11.9905	3	3	−0.2
7	9.3973	4	4	0.0
8	9.3513	4	4	−0.0

N	SV	W_p	W_r	PD (deg.)
1	19.2326	1	1	−3.3
2	17.7614	1	1	−3.2
3	16.6002	2	2	−5.8
4	15.5774	2	2	2.9
5	13.2590	3	3	−8.9
6	12.7727	3	3	−9.0
7	10.7077	4	4	9.1
8	10.3162	4	4	9.2

the difference between the magnitudes of singular values. The singular vectors for both ideal and FDT operators fall onto wavenumbers 1, 2, 3 and 4. The phase difference here is captured rather well and it is essentially zero in all cases.

2.6.2. The validity of FDT for K-Z model. As we did above for TBH, first we report the climatology for K-Z model. The variance spectrum of the K-Z model as defined above in (2.86) is shown in Figure 2.11. The variance spectrum in Figure 2.11 is uniform as predicted by (2.78).

The time autocorrelation functions are shown in Figure 2.12 for the wavenumbers $k = 1, \ldots, 4$. A general formula for the autocorrelation function of a variable $a(t)$ is given in (2.87) as convolution of a wavenumber in time. In practice the time averaging window T is finite but large (10^5 time units in our simulations). The autocorrelation function in (2.87) measures mixing in a dynamical system, and in

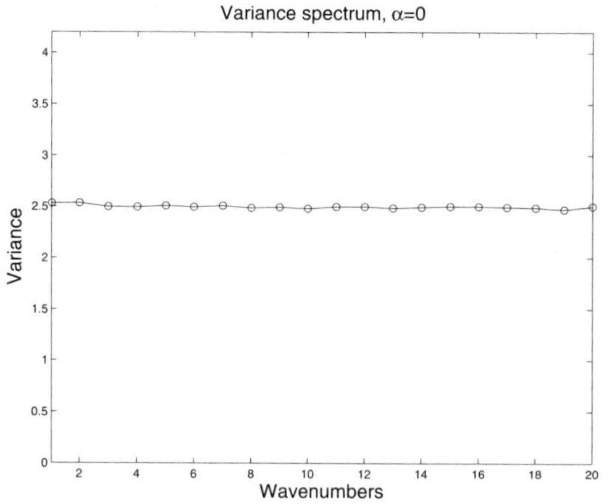

FIGURE 2.11. The variance spectrum for the Kruskal–Zabusky model.

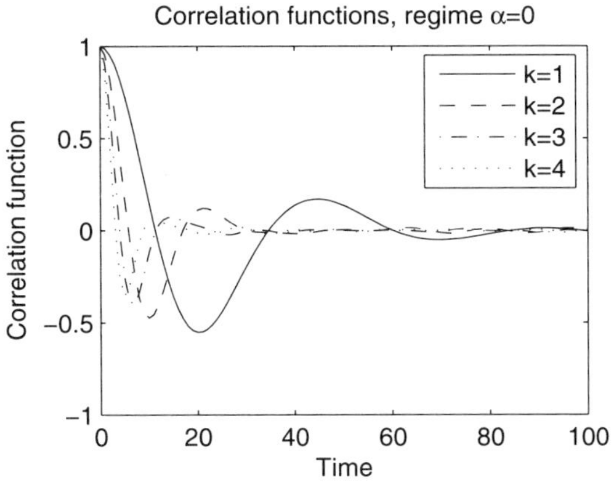

FIGURE 2.12. The time autocorrelation functions for the wavenumbers $k = 1, \ldots, 4$, Kruskal–Zabusky model. Solid line—$k = 1$, dashed line—$k = 2$, dot-dashed line—$k = 3$, dotted line—$k = 4$.

some sense can be regarded as an indicator of how fast a variable "forgets" about its behavior in the past. A distinct scaling for correlation function behavior can be observed as mentioned in (2.82)—that is, the time autocorrelation functions evolve slowly for large scale wavenumbers, whereas mid-scale wavenumbers mix faster. It is, however, remarkable that small scale wavenumbers in the K-Z model mix moderately fast—that is, faster than large scales, but slower than mid-scales. This feature of the K-Z model is clearly seen on the autocorrelation time plots which are presented below.

The correlation time is defined as a typical time of the decay of the autocorrelation function. Usually it is given through the time integral of the autocorrelation function. Here we look at two different measures of the correlation time: one is the integral of the autocorrelation function in (2.88), and the other is the integral of the absolute value of the autocorrelation function in (2.89). Clearly $T_{\text{abs}} \geq T$ and T_{abs} accounts for all the oscillations in the correlation function. As for the TBH model, here the two types of correlation time spectra are presented in Figure 2.13: the correlation times shown by solid lines in Figure 2.13 are obtained by integrating the time autocorrelation functions, whereas the correlation times shown by dashed lines in Figure 2.13 are obtained by integrating the absolute value of the corresponding correlation function. Note that the T and T_{abs} differ significantly at large scales, moderately at mid-scales, and not at all at small scales. This means that negative portions of time autocorrelation function for the K-Z model almost cancel out the positive parts at large scales, with very little oscillation in the correlation functions at small scales. We also observe that, contrary to the TBH model, here the small scale wavenumbers have moderately decaying correlations compared to mid-scale wavenumbers.

Now, we return to the validity of FDT for the K-Z model. As in case with TBH, here the ideal linear response operator is constructed for the simplest functional $\vec{A}(\vec{x}) = \vec{x}$. The FDT linear response operator is constructed in Fourier space.

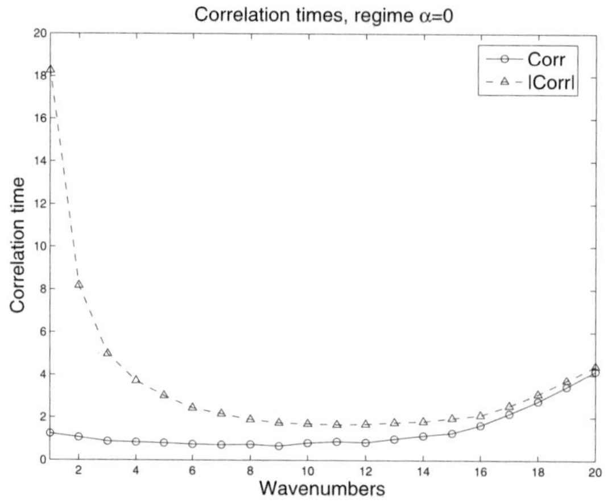

FIGURE 2.13. The spectra of correlation times (integrals of correlation functions) for the Kruskal–Zabusky model. Solid line with circles: the integrals are taken with the time correlation functions directly. Dashed line with triangles: the integrals are taken with the absolute values of the correlation functions.

Further we do not show the errors $K(k, r, t)$ since they are comparable in magnitude with those in Figures 2.7 and 2.10 for the Burgers–Hopf model.

2.6.2.1. *Time evolution of singular values.* The time evolution of the singular values 1–4 for the K-Z model is shown in Figure 2.14. Each picture contains two curves: solid—the result for the ideal operator, and dashed—the result for the FDT operator. One can observe that the evolution or the ideal and FDT singular values have roughly the same qualitative behavior. First there is a stage where the ideal and FDT results coincide with each other (roughly until the time $T = 5$), and later the FDT predictions generally overshoot the ideal measurements. The kinks in these graphs represent bifurcations in time where different spatial wavenumbers control the largest response. Below we study the linear response in detail for three different time snapshots. What we will observe is that for the early time $T = 5$ the response is concentrated on large scales, at time $T = 10$ the response is shared between large scales and small scales with the gap at mid-scales, and at the time $T = 30$ the response moves entirely to small scales.

2.6.2.2. *The response operators for time $T = 5$.* Here we show more extensive results for the time slice $T = 5$. In Figure 2.15 we display the ideal response operators and the FDT approximations in physical space, with their main diagonals in the Fourier space. Like for the TBH model, here the general response structure in the Fourier space is also concentrated on the main diagonal. This is common for both ideal and approximate FDT operators. Observe that the bulk of the response is shared between large and small scale modes, leaving almost nothing in between. The linear response in physical space is concentrated at the gridpoint of perturbation, however there are also oscillations around it. We can see little differences between the ideal operator and the FDT operator, which means that the FDT framework is valid for the K-Z model. The spectra of singular values for the time $T = 5$ are shown in Figure 2.16, which contains two curves: solid curve

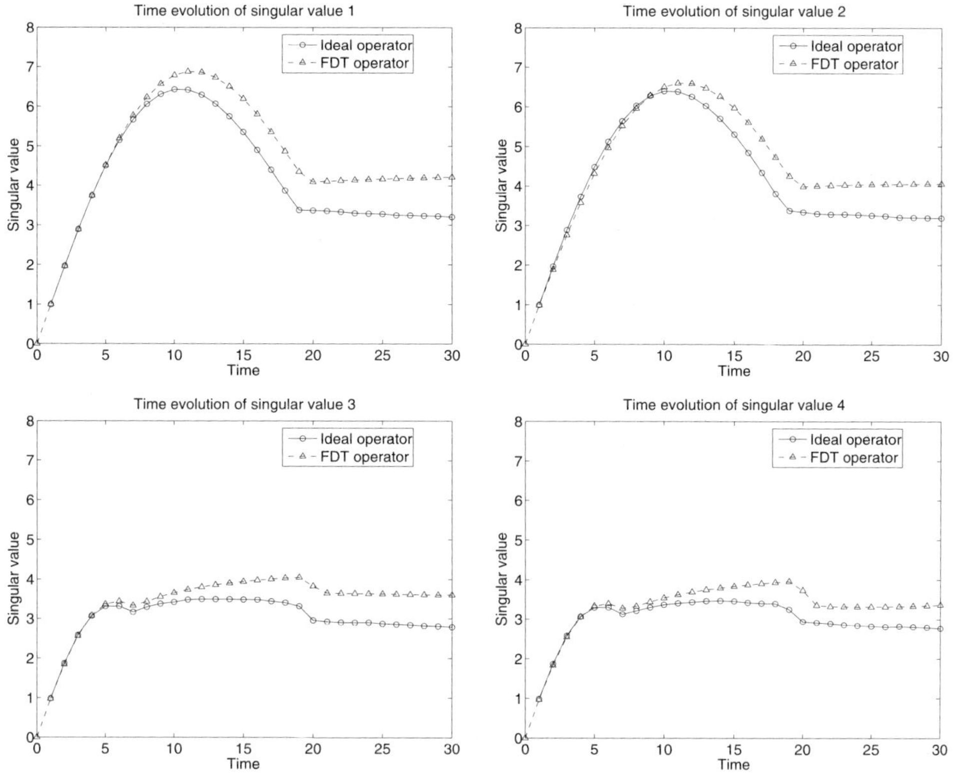

FIGURE 2.14. The time evolution of singular values 1, 2, 3 and 4 for the Kruskal–Zabusky model. Solid line—singular values of the ideal linear response operator. Dashed line—the singular values of the FDT response operators.

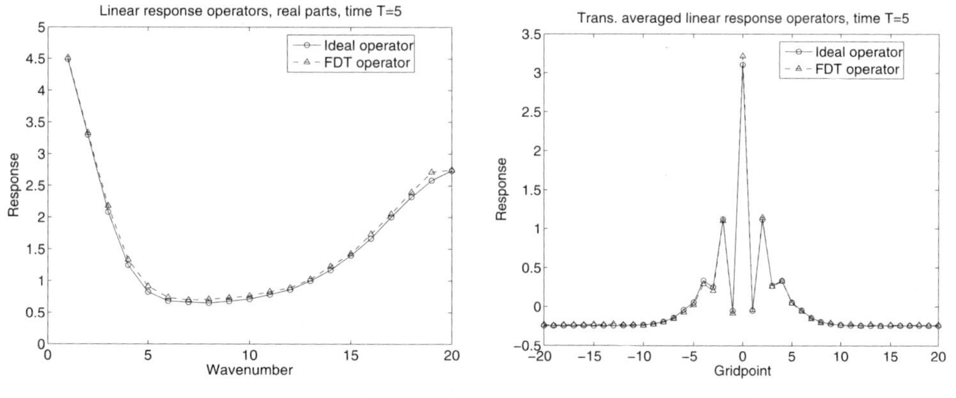

FIGURE 2.15. Left—real part of the main diagonal of linear response operator in Fourier space. Right—Translationally averaged linear response operators, in physical space. Solid line—ideal operator, dashed line—FDT operator. Time $T = 5$, the Kruskal–Zabusky model.

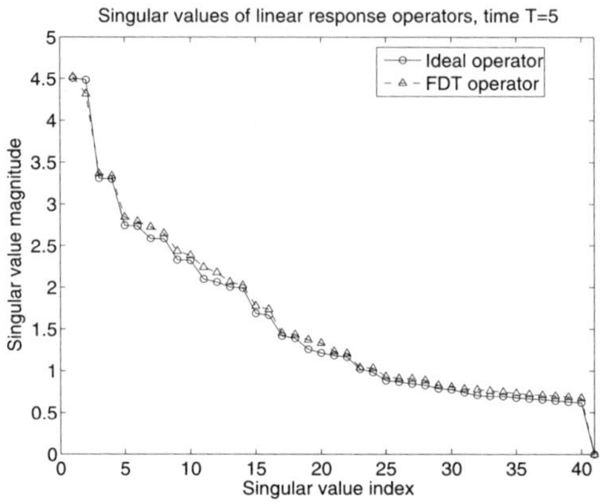

FIGURE 2.16. The spectrum of singular values for linear response operators. Solid line—ideal operator, dashed line—FDT operator. Time $T = 5$, the Kruskal–Zabusky model.

TABLE 2.5. The errors and correlations for linear response operators at time $T = 10$ for the Burgers–Hopf model.

	L^2 error	$C(g,h)$
Fourier diag.	0.05398	0.9993
Phys. trans. av.	0.115	0.9943

TABLE 2.6. The errors and correlations for linear response operators at time $T = 5$ for the Kruskal–Zabusky model.

	L^2 error	$C(g,h)$
Fourier diag.	0.0326	0.9998
Phys. trans. av.	0.04057	0.9994

denotes the results for the ideal linear response operator obtained by perturbing the system and measuring the response; the dashed curve shows the results for the FDT approximation of the response operator. As we can see, the FDT spectra of singular values capture the actual spectrum moderately well.

In Table 2.6 we show the errors (2.84) and correlations (2.85) for the time $T = 5$. The errors and correlations are presented for the real and imaginary parts of main diagonals of linear response operators in the Fourier space, for the translationally averaged linear response operators in physical space, and for the spectra of singular values. Since the errors are small for all measurements, the ideal and FDT results are expected to be collinear, and Table 2.6 confirms this fact. In Table 2.7 we summarize the results for the time $T = 5$. These two tables correspond to the ideal and FDT results, respectively. Each table contains the following fields: SW—contains singular values, W_p—the wavenumber of corresponding singular perturbation, W_r—the wavenumber of corresponding singular response, and PD—the phase difference between the perturbation and the response. As we can

TABLE 2.7. The overview of results at time $T = 5$, first 8 singular values. Left—ideal operator, right—FDT. Notations: SV—singular values, W_p—wavenumber on which singular perturbation is projected, W_r—wavenumber on which singular response is projected, PD—phase difference between singular perturbation and singular response. The Kruskal–Zabusky model.

N	SV	W_p	W_r	PD (deg.)
1	4.5046	1	1	−0.0
2	4.4831	1	1	−0.0
3	3.3096	2	2	−0.0
4	3.3011	2	2	0.1
5	2.7452	20	20	−0.1
6	2.7369	20	20	0.1
7	2.5870	19	19	−0.3
8	2.5861	19	19	0.3

N	SV	W_p	W_r	PD (deg.)
1	4.5192	1	1	0.5
2	4.3166	1	1	0.5
3	3.3658	2	2	1.6
4	3.3380	2	2	1.6
5	2.8453	20	20	0.3
6	2.7933	19	19	−0.4
7	2.7261	20	20	0.4
8	2.6491	19	19	−0.3

see, all wavenumbers are captured perfectly by the FDT method, which is surprising given the difference between the magnitudes of singular values. The singular vectors for both ideal and FDT operators fall onto wavenumbers 1, 2, 20 and 19. Note the remarkable difference from the TBH model, where largest responses are always observed at large scales. The phase difference here is captured rather well and it is essentially zero in all cases.

2.6.2.3. *The response operators for time $T = 10$.* Here we show more extensive results for the time slice $T = 10$ for the K-Z model. In Figure 2.17 we display the ideal response operators and the FDT approximations in physical space, with their main diagonals in the Fourier space. Like for the time $T = 5$, the general response structure in the Fourier space is concentrated on the main diagonal. This is common

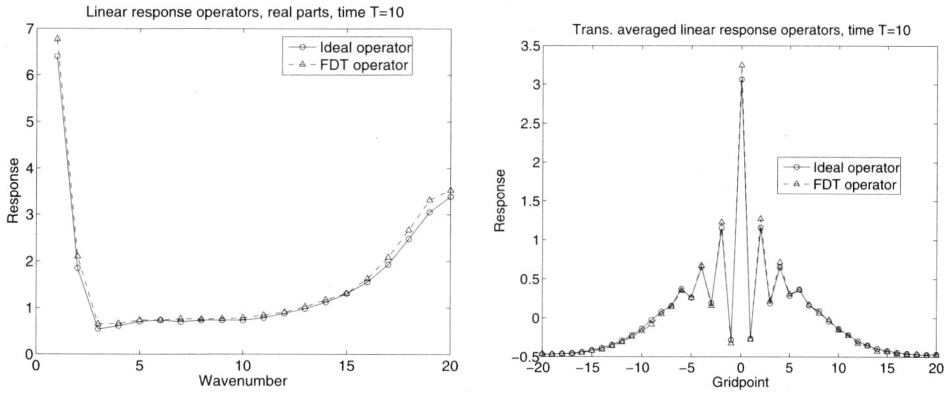

FIGURE 2.17. Left—real part of the main diagonal of linear response operator in Fourier space. Right—Translationally averaged linear response operators, in physical space. Solid line—ideal operator, dashed line—FDT operator. Time $T = 10$, the Kruskal–Zabusky model.

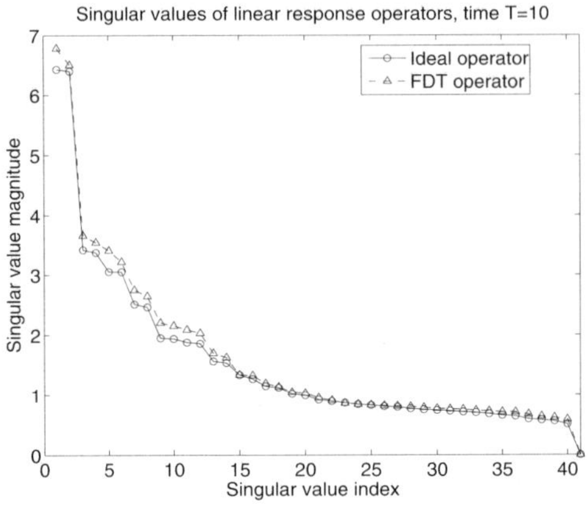

FIGURE 2.18. The spectrum of singular values for linear response operators. Solid line—ideal operator, dashed line—FDT operator. Time $T = 10$, the Kruskal–Zabusky model.

for both ideal and approximate FDT operators. For the physical space response, the oscillations are somewhat stronger around the peak of the main response, if compared to the previous time $T = 5$. We can see little differences between the ideal operator and the FDT operator, which means that the FDT framework is valid for the current model. The spectra of singular values for the time $T = 10$ are shown in Figure 2.18, which contains two curves: solid curve denotes the results for the ideal linear response operator obtained by perturbing the system and measuring the response; dashed curve shows the results for the FDT approximation of the response operator. As we can see, the FDT spectra of singular values capture the actual spectrum with minor discrepancies.

In Table 2.5 we show the errors (2.84) and correlations (2.85) for the time $T = 10$. The errors and correlations are presented for the real and imaginary parts of the main diagonals of linear response operators in the Fourier space, for the translationally averaged linear response operators in physical space, and for the spectra of singular values. The errors are not as small as for the time $T = 5$, however the ideal and FDT results are still collinear. In Table 2.8 we summarize the results for the time $T = 10$. These two tables correspond to the ideal and FDT results, respectively. Each table contains the following fields: SW—contains singular values, W_p—the wavenumber of corresponding singular perturbation, W_r—the wavenumber of corresponding singular response, and PD—the phase difference between the perturbation and the response. As we can see, all wavenumbers are captured perfectly by the FDT method, which is surprising given the difference between the magnitudes of singular values. The singular vectors for the ideal operator fall onto wavenumbers 1, 20, 19 and 2, whereas the results for the FDT operator fall onto wavenumbers 1, 20, 19 and 18. Again observe the difference between the TBH and K-Z models: here the largest responses are rather evenly split between the largest and smallest scales, whereas for the TBH model the largest responses are recorded at largest scales. The phase difference here is captured rather well and it is essentially zero in all cases.

TABLE 2.8. The overview of results at time $T = 10$, first 8 singular values. Left—ideal operator, right—FDT. Notations: SV—singular values, W_p—wavenumber on which singular perturbation is projected, W_r—wavenumber on which singular response is projected, PD—phase difference between singular perturbation and singular response. The Kruskal–Zabusky model.

N	SV	W_p	W_r	PD (deg.)
1	6.4285	1	1	0.0
2	6.3977	1	1	0.0
3	3.4189	20	20	−0.1
4	3.3726	20	20	0.1
5	3.0557	19	19	−0.1
6	3.0512	19	19	0.1
7	2.5139	18	18	−0.4
8	2.4654	18	18	0.3

N	SV	W_p	W_r	PD (deg.)
1	6.7853	1	1	0.9
2	6.5011	1	1	0.9
3	3.6578	20	20	0.7
4	3.5420	20	20	−0.7
5	3.4069	19	19	−0.7
6	3.2192	19	19	−0.8
7	2.7510	18	18	0.0
8	2.6480	18	18	−0.0

TABLE 2.9. The errors and correlations for linear response operators at time $T = 30$ for the Kruskal–Zabusky model.

	L^2 error	$C(g,h)$
Fourier diag.	0.1958	0.9939
Phys. trans. av.	0.2045	0.9924

2.6.2.4. *The response operators for time* $T = 30$. Here we show more extensive results for the time slice $T = 30$. In Figure 2.19 we display the ideal response operators and the FDT approximations in physical space, with their main diagonals

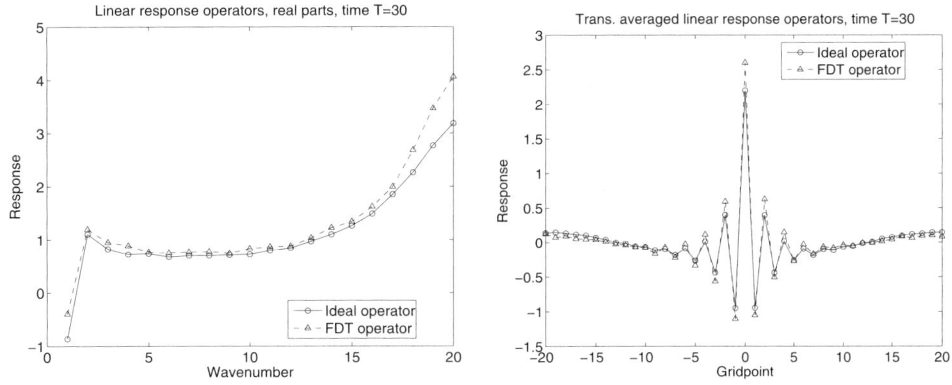

FIGURE 2.19. Left—real part of the main diagonal of linear response operator in Fourier space. Right—Translationally averaged linear response operators, in physical space. Solid line—ideal operator, dashed line—FDT operator. Time $T = 30$, the Kruskal–Zabusky model.

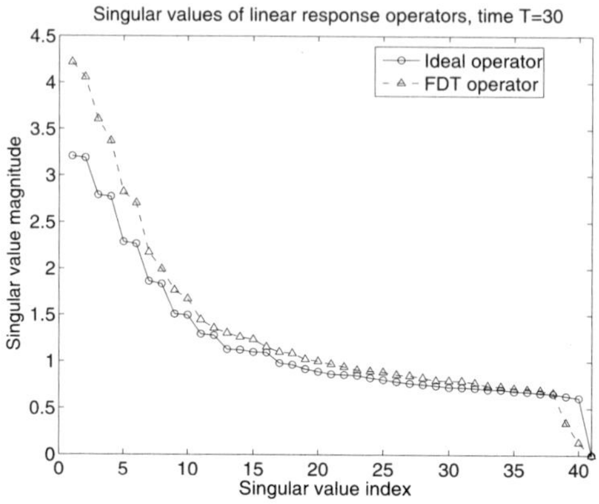

FIGURE 2.20. The spectrum of singular values for linear response operators. Solid line—ideal operator, dashed line—FDT operator. Time $T = 30$, the Kruskal–Zabusky model.

in the Fourier space. The general response structure in the Fourier space is concentrated on the main diagonal. This is common for both ideal and approximate FDT operators. The new remarkable feature here is that, unlike the TBH model and the K-Z model at earlier times, all responses are concentrated on the smallest scales, and yet the FDT framework is able to capture them. We can see almost no differences between the ideal operator and the FDT operator, which means that the FDT framework is valid for the current model. The spectra of singular values for the time $T = 30$ are shown in Figure 2.20, which contains two curves: solid curve denotes the results for the actual (ideal) linear response operator obtained by perturbing the system and measuring the response; dashed curve shows the results for the FDT approximation of the response operator. As we can see, the FDT spectra of singular values capture the actual spectrum almost precisely.

TABLE 2.10. The overview of results at time $T = 30$, first 8 singular values. Left—ideal operator, right—FDT. Notations: SV—singular values, W_p—wavenumber on which singular perturbation is projected, W_r—wavenumber on which singular response is projected, PD—phase difference between singular perturbation and singular response. The Kruskal–Zabusky model.

N	SV	W_p	W_r	PD (deg.)
1	3.2061	20	20	−0.1
2	3.1879	20	20	0.1
3	2.7908	19	19	−0.2
4	2.7730	19	19	0.1
5	2.2868	18	18	0.0
6	2.2662	18	18	−0.1
7	1.8642	17	17	−0.2
8	1.8368	17	17	0.2

N	SV	W_p	W_r	PD (deg.)
1	4.2175	20	20	1.0
2	4.0570	20	20	0.7
3	3.6082	19	19	−1.1
4	3.3681	19	19	−0.7
5	2.8219	18	18	−1.8
6	2.7041	18	18	−1.8
7	2.1749	17	17	−1.3
8	1.9947	17	17	−2.0

In Table 2.9 we show the errors (2.84) and correlations (2.85) for the time $T = 30$. The errors and correlations are presented for the real and imaginary parts of main diagonals of linear response operators in the Fourier space, for the translationally averaged linear response operators in physical space, and for the spectra of singular values. It is interesting that, even though the errors now are larger than for previous times, the FDT predictions and ideal measurements are still very collinear. In Table 2.10 we summarize the results for the K-Z model at the time $T = 30$. These two tables correspond to the ideal and FDT results, respectively. Each table contains the following fields: SW—contains singular values, W_p—the wavenumber of corresponding singular perturbation, W_r – the wavenumber of corresponding singular response, and PD—the phase difference between the perturbation and the response. As we can see, all wavenumbers are captured perfectly by the FDT method, which is surprising given the difference between the magnitudes of singular values. The singular vectors for the ideal and FDT operators fall onto wavenumbers 20, 19, 18 and 17. Here observe that the largest responses are concentrated on small scales (direct consequence of the correlation time scaling in Figure 2.13), which is the opposite behavior to that of the TBH model, which responds mainly on large scales. Yet the FDT proves to be valid in this rather extraordinary situation. The phase difference here is captured rather well and it is essentially zero in all cases.

2.6.3. The validity of FDT for IL96 model. As we did above for the TBH and K-Z models, first we report the climatology for IL96 model. The variance spectrum of the IL96 model as defined above in (2.86) is shown in Figure 2.21. The variance spectrum in Figure 2.21 is uniform as predicted by (2.78).

The time autocorrelation functions are shown in Figure 2.22 for the wavenumbers $k = 1, \ldots, 4$. A general formula for the autocorrelation function of a variable $a(t)$ is given in (2.87) as convolution of a wavenumber in time. In practice the time averaging window T is finite but large (10^5 time units in our simulations). The autocorrelation function in (2.87) measures mixing in a dynamical system, and in some sense can be regarded as an indicator of how fast a variable "forgets" about its behavior in the past. Neither correlation time scaling nor oscillations are observed

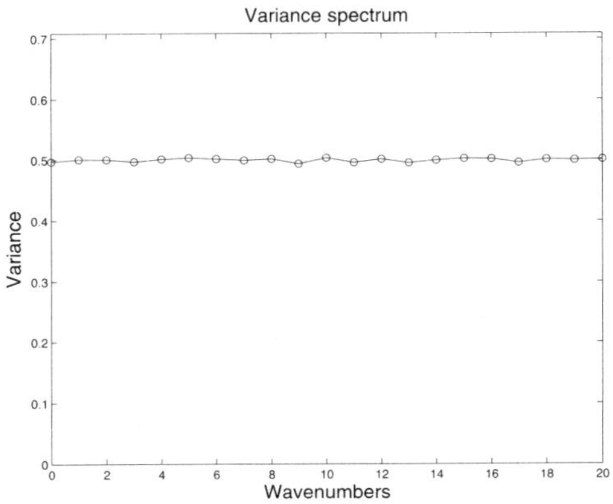

FIGURE 2.21. The variance spectrum for the IL96 model.

here as was also mentioned in (2.82), in fact, all the correlation functions decay with a similar rate.

The correlation time is defined as a typical time of the decay of the autocorrelation function. Usually it is given through the time integral of the autocorrelation function. Here we look at two different measures of the correlation time: one is the integral of the autocorrelation function in (2.88), and the other is the integral of the absolute value of the autocorrelation function in (2.89). Clearly $T_{\text{abs}} \geq T$ and T_{abs} accounts for all the oscillations in the correlation function. As for the TBH and K-Z models, here the two types of correlation time spectra are presented in Figure 2.23: the correlation times shown by solid lines in Figure 2.23 are obtained by integrating the time autocorrelation functions, whereas the correlation times shown

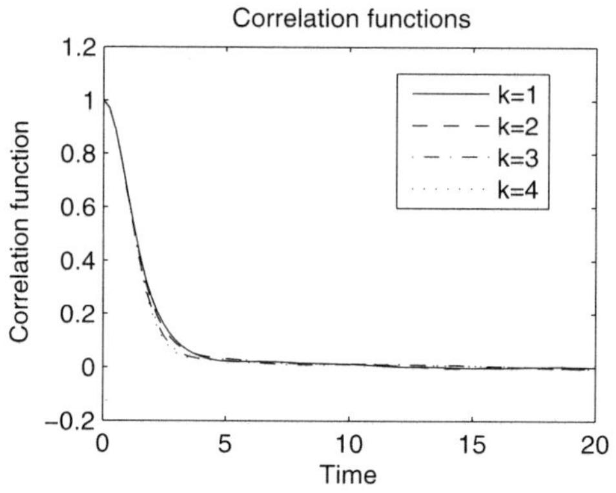

FIGURE 2.22. The time autocorrelation functions for the wavenumbers $k = 1, \ldots, 4$, IL96 model. Solid line—$k = 1$, dashed line – $k = 2$, dot-dashed line—$k = 3$, dotted line—$k = 4$.

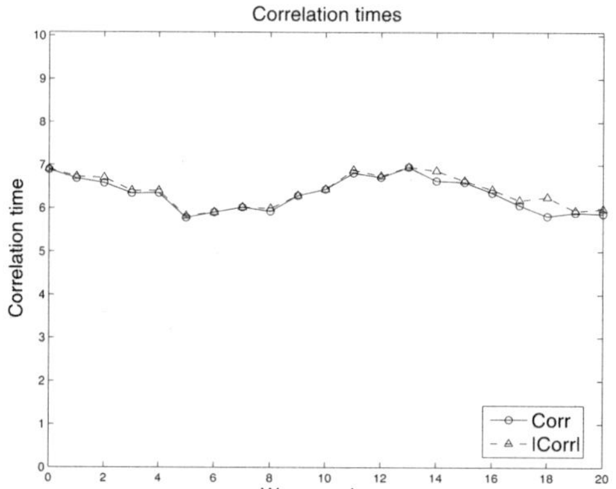

FIGURE 2.23. The spectra of correlation times (integrals of correlation functions) for the IL96 model. Solid line with circles: the integrals are taken with the time correlation functions directly. Dashed line with triangles: the integrals are taken with the absolute values of the correlation functions.

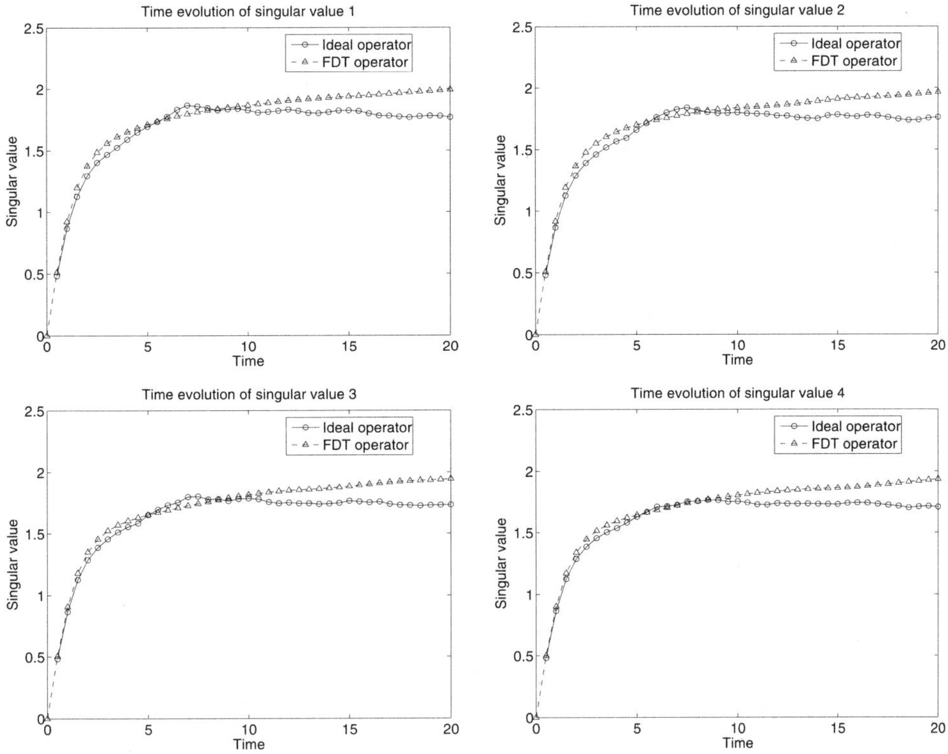

FIGURE 2.24. The time evolution of singular values 1, 2, 3 and 4 for the IL96 model. Solid line—singular values of the ideal linear response operator. Dashed line—the singular values of the FDT response operators.

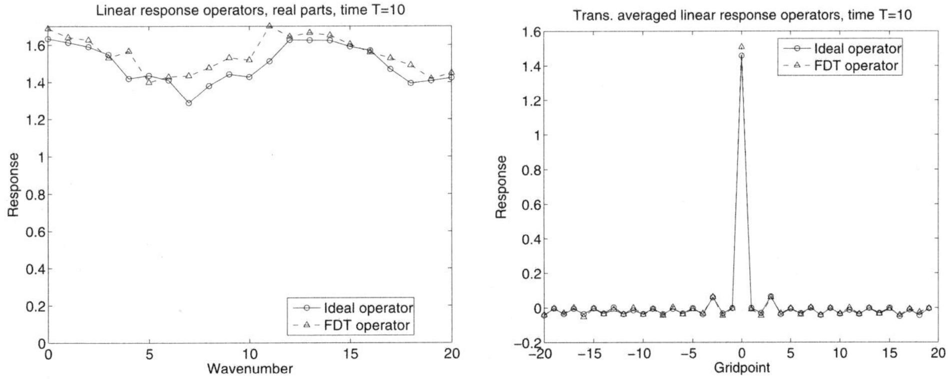

FIGURE 2.25. Left—real part of the main diagonal of linear response operator in Fourier space. Right—Translationally averaged linear response operators, in physical space. Solid line—ideal operator, dashed line—FDT operator. Time $T = 10$, the IL96 model.

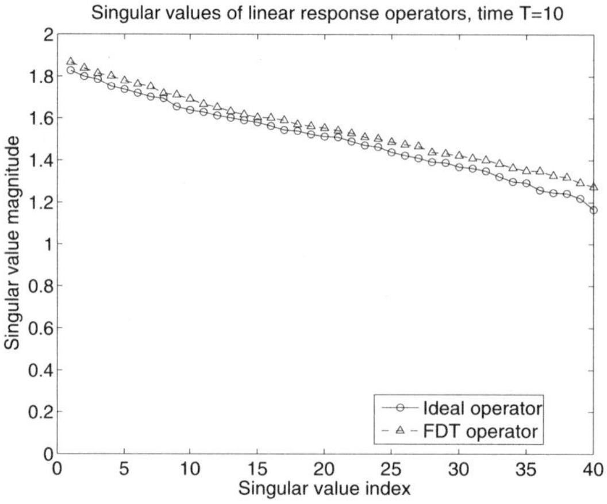

FIGURE 2.26. The spectrum of singular values for linear response operators. Solid line—ideal operator, dashed line—FDT operator. Time $T = 10$, the IL96 model.

by dashed lines in Figure 2.23 are obtained by integrating the absolute value of the corresponding correlation function. First, observe that there is indeed no scaling of autocorrelation times, in contrast to TBH and K-Z models. Second, note that the T and T_{abs} do not differ significantly from each other, which means that negative portions of time autocorrelation functions for the IL96 model are negligible for all wavenumbers as noted earlier in (2.82).

Now, we return to the validity of FDT for the IL96 model. As in the case with TBH and K-Z models, here the ideal linear response operator is constructed for the simplest functional $\vec{A}(\vec{x}) = \vec{x}$. The FDT linear response operator is constructed in Fourier space.

2.6.3.1. *Time evolution of singular values.* The time evolution of the singular values 1–4 for the IL96 model is shown in Figure 2.24. Each picture contains two curves: solid—the result for the ideal operator, and dashed—the result for the FDT operator. One can observe that the evolution of the ideal and FDT singular values have roughly the same qualitative behavior. First there is a stage where the ideal and FDT results coincide with each other (roughly until the time $T = 10$), and later the FDT results generally overshoot the ideal results. There is almost no difference in magnitudes of different singular values, which is a direct consequence of the flat correlation time graph shown in Figure 2.23. Below we study the response in detail at two different time slices, $T = 10$ (this is where the FDT and ideal singular values start diverging), and $T = 20$.

2.6.3.2. *The response operators for time $T = 10$.* Here we show more extensive results for the time slice $T = 10$ for the K-Z model. In Figure 2.25 we display the ideal response operators and the FDT approximations in physical space, with their main diagonals in Fourier space. As for the TBH and K-Z models, the general response structure in the Fourier space is concentrated on the main diagonal, but we do not show the imaginary part of the main diagonal due to its similarity with the real part. This is common for both ideal and approximate FDT operators. The response in physical space shows a peak at the perturbation gridpoint and

TABLE 2.11. The errors and correlations for linear response operators at time $T = 10$ for the IL96 model.

	L^2 error	$C(g, h)$
Fourier diag.	0.05095	0.9993
Phys. trans. av.	0.04591	0.9995

the small scale response elsewhere. We can see little differences between the ideal operator and the FDT operator, which means that the FDT framework is valid for the current model. The spectra of singular values for the time $T = 10$ are shown in Figure 2.26, which contains two curves: solid curve denotes the results for the ideal linear response operator obtained by perturbing the system and measuring the response; dashed curve shows the results for the FDT approximation of the response operator. As we can see, the FDT spectra of singular values capture the actual spectrum with minor discrepancies.

In Table 2.11 we show the errors (2.84) and correlations (2.85) for the time $T = 10$. The errors and correlations are presented for the real and imaginary parts of main diagonals of linear response operators in the Fourier space, for the translationally averaged linear response operators in physical space, and for the spectra of singular values. Observe that the errors are small and the FDT predictions and ideal measurements are collinear. In Table 2.12 we summarize the results for the time $T = 10$. These two tables correspond to the ideal and FDT results, respectively. Each table contains the following fields: SW – contains singular values, W_p—the wavenumber of corresponding singular perturbation, W_r—the wavenumber of corresponding singular response, and PD—the phase difference between the perturbation and the response. Observe that the FDT fails completely to capture correct wavenumbers here, mainly due to the fact that largest and smallest singular values are very close to each other, so that even a small error induces such qualitative change in the wavenumber projection, but these errors are not significant here. This is a distinct difference in behavior of the IL96 model from TBH and K-Z models, which is also confirmed in the flat correlation time plot in Figure 2.23. However, regardless of whether the projection wavenumbers are captured correctly by FDT or not, the errors in linear responses are small (as we saw already), thus we see that FDT is valid for the IL96 model.

2.6.3.3. *The response operators for time $T = 20$.* Here we show more extensive results for the time slice $T = 20$ for K-Z model. In Figure 2.27 we display the ideal response operators and the FDT approximations in physical space, with their main diagonals in the Fourier space. The general response structure in the Fourier space is concentrated on the main diagonal. This is common for both ideal and approximate FDT operators. We can see little differences between the ideal operator and the FDT operator, which means that the FDT framework is valid for the current model. The spectra of singular values for the time $T = 20$ are shown in Figure 2.28, which contains two curves: solid curve denotes the results for the ideal linear response operator obtained by perturbing the system and measuring the response; dashed curve shows the results for the FDT approximation of the response operator. As we can see, the FDT spectra of singular values capture the actual spectrum with minor discrepancies.

TABLE 2.12. The overview of results at time $T = 10$, first 8 singular values. Left—ideal operator, right—FDT. Notations: SV—singular values, W_p—wavenumber on which singular perturbation is projected, W_r—wavenumber on which singular response is projected, PD—phase difference between singular perturbation and singular response. The IL96 model.

N	SV	W_p	W_r	PD (deg.)
1	1.8273	1	1	−0.5
2	1.7996	0	0	0.0
3	1.7852	14	14	−0.6
4	1.7514	12	12	−0.7
5	1.7368	1	1	0.9
6	1.7192	12	12	−0.7
7	1.7011	2	2	−0.0
8	1.6935	14	13	133.5

N	SV	W_p	W_r	PD (deg.)
1	1.8674	13	13	−0.0
2	1.8390	13	13	3.2
3	1.8142	11	11	1.3
4	1.8014	12	12	0.3
5	1.7781	12	12	1.8
6	1.7624	1	1	−2.4
7	1.7502	1	1	2.1
8	1.7170	16	12	148.6

TABLE 2.13. The errors and correlations for linear response operators at time $T = 20$ for the IL96 model.

	L^2 error	$C(g, h)$
Fourier diag.	0.06545	0.9991
Phys. trans. av.	0.06136	0.9993

In Table 2.13 we show the errors (2.84) and correlations (2.85) for the time $T = 20$. The errors and correlations are presented for the real and imaginary parts of the main diagonals of linear response operators in the Fourier space, for the translationally averaged linear response operators in physical space, and for the spectra of singular values. The errors here are somewhat larger than those for the time $T = 10$, however the FDT predictions and ideal measurements are still collinear.

In Table 2.14 we summarize the results for the time $T = 20$. These two tables correspond to the ideal and FDT results, respectively. Each table contains the following fields: SW—contains singular values, W_p—the wavenumber of corresponding singular perturbation, W_r—the wavenumber of corresponding singular response, and PD—the phase difference between the perturbation and the response. Like for the time $T = 10$, observe that here the FDT fails completely to capture correct wavenumbers, mainly due to the fact that largest and smallest singular values are very close to each other, so that even a small error induces such qualitative change in the wavenumber projection. However, regardless of whether the projection wavenumbers are captured correctly by FDT or not, the errors in linear responses are small at time $T = 20$, which means that the FDT is valid.

2.6.4. Summary. Here we discuss the results of the FDT approximation for TBH, K-Z and IL96. Overall the FDT approximation for these models is outstanding for the time scales considered. The three models, although being truncations of the inviscid Burgers–Hopf equation with the same equipartition spectrum, nevertheless have different time correlation scaling and thus different properties of the

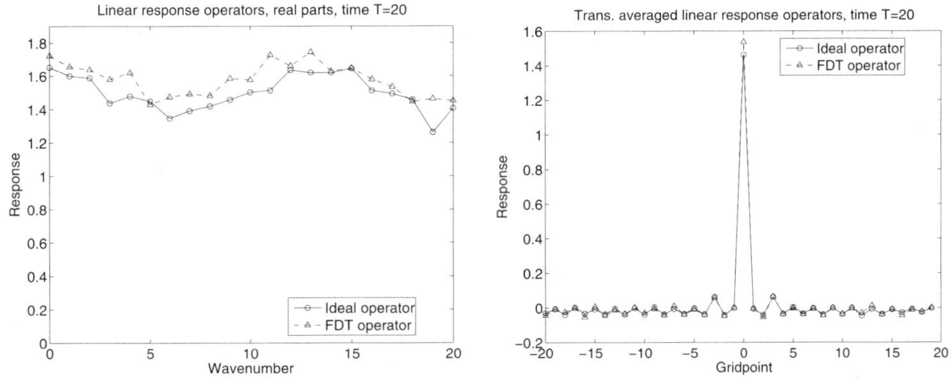

FIGURE 2.27. Left—real part of the main diagonal of linear response operator in Fourier space. Right—Translationally averaged linear response operators, in physical space. Solid line—ideal operator, dashed line—FDT operator. Time $T = 20$, the IL96 model.

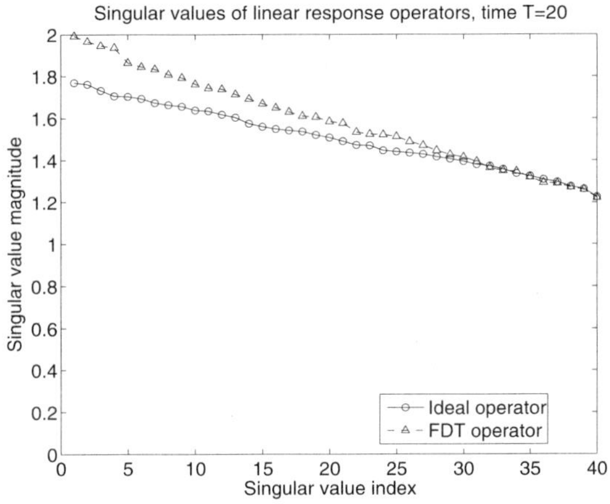

FIGURE 2.28. The spectrum of singular values for linear response operators. Solid line—ideal operator, dashed line—FDT operator. Time $T = 20$, the IL96 model.

response to small forcing. The TBH model is characterized by a correlation time scaling inversely proportional to wavenumber index, which means that the response is stronger at large scales, and weaker at small scales at all times. The K-Z model represents a situation where for short times the response is like that for the TBH model, for moderate times it is divided between large scales and small scales with essentially no response at mid-scales, and for long times the response shifts completely to small scales. The IL96 model is a system with no correlation scaling, that is, all the correlation functions have a similar decay rate. Thus, the three models span a wide range of potential physical situations with response at vastly different scales. The FDT is observed to be valid for all three models, which suggests its practical applicability for various qualitatively different situations. In particular,

TABLE 2.14. The overview of results at time $T = 20$, first 8 singular values. Left—ideal operator, right—FDT. Notations: SV—singular values, W_p—wavenumber on which singular perturbation is projected, W_r—wavenumber on which singular response is projected, PD—phase difference between singular perturbation and singular response. The IL96 model.

N	SV	W_p	W_r	PD (deg.)
1	1.7692	14	14	−1.0
2	1.7619	0	0	0.0
3	1.7321	12	12	1.4
4	1.7059	13	13	0.1
5	1.7036	2	2	−0.2
6	1.6934	14	14	−1.9
7	1.6734	1	1	3.1
8	1.6633	12	12	1.4

N	SV	W_p	W_r	PD (deg.)
1	1.9926	11	11	−0.3
2	1.9657	13	13	1.0
3	1.9451	13	13	−0.2
4	1.9352	13	13	−1.7
5	1.8641	12	12	6.3
6	1.8446	16	16	−0.5
7	1.8332	4	4	1.2
8	1.8076	0	0	0.0

the evolution of singular values, the time snapshots of linear response operators and the projection wavenumbers are captured perfectly for all observed times (the latter apparently due to distinct correlation time scaling). For the K-Z and IL96 model, there is an initial interval of time when the evolution of singular values is indistinguishable from the ideal response, however even after that time the FDT and ideal singular values are very near. The profiles of the response operators are captured very well in both physical and Fourier spaces. The transition of the response from large scales to small scales for K-Z model is captured well by the FDT approximation. For the IL96 model, the absence of the correlation time scaling does not allow us to capture the correct order of the projected wavenumbers of the response, but the profile of the linear response operator in both physical and Fourier spaces are also captured very well by the FDT.

2.7. The quasi-Gaussian approximation for FDT in forced, damped nonlinear systems

In Section 2.6 we have demonstrated the remarkable accuracy of the predictions of FDT for three diverse nonlinear dynamical systems without damping and forcing but with many degrees of freedom and a Gaussian invariant measure. Most practical nonlinear systems in the geosciences also involve significant damping and forcing. If there is no random forcing, probability densities for statistical solutions evolve and solve the general Liouville operator in (2.4). In its contemporary use for deterministic damped forced systems without stochastic noise, formulas from Theorem 2.1 are utilized as derived in Section 2.2 under the approximation that the invariant measure $p_{\text{eq}}(\vec{X})$ is replaced by $p^G_{\text{eq}}(\vec{X})$, the Gaussian measure with the same mean and second moments; in other words, p^G_{eq} is given explicitly by

$$p^G_{\text{eq}}(\vec{X}) = (2\pi)^{-N/2}(\det C)^{-1/2}\exp\left(-\tfrac{1}{2}(\vec{X}-\bar{\vec{X}})^T C^{-1}(\vec{X}-\bar{\vec{X}})\right) \tag{2.90}$$

where the mean $\bar{\vec{X}}$ and the positive definite symmetric covariance matrix C for p_{eq}^G are chosen to match those for p_{eq},

$$\int \vec{X} p_{\text{eq}}^G(\vec{X})\, d\vec{X} = \bar{\vec{X}} = \int \vec{X} p_{\text{eq}}(\vec{X})\, d\vec{X},$$

$$\int (\vec{X} - \bar{\vec{X}}) \otimes (\vec{X} - \bar{\vec{X}}) p_{\text{eq}}^G(\vec{X})\, d\vec{X} = C = \int (\vec{X} - \bar{\vec{X}}) \otimes (\vec{X} - \bar{\vec{X}}) p_{\text{eq}}(\vec{X})\, d\vec{X}. \tag{2.91}$$

An approximation, $\vec{R}^G(\tau)$, to the response operator, $\vec{R}(\tau)$, in Theorem 2.1 from Section 2.2 is made through the formula

$$\vec{R}^G(\tau) = \langle A(\vec{X}(t+\tau)) \vec{B}^G(\vec{X}(t)) \rangle \tag{2.92}$$

with

$$\vec{B}^G(\vec{X}) = \frac{\vec{L}_{\text{ext}} p_{\text{eq}}^G}{p_{\text{eq}}^G}. \tag{2.93}$$

Note that although the Gaussian approximation defined in (2.90), (2.91) is utilized in the approximate linear response operator in (2.92), the average in brackets actually occurs with respect to the original invariant measure, $p_{\text{eq}}(\vec{X})$. This is especially evident for a strongly mixing dissipative dynamical system where, as discussed in Sections 2.5 and 2.6, the correlation function in (2.92) is calculated by a long time average over a single trajectory. This is why we use the term "quasi-Gaussian approximation for FDT" although Dymnikov and Gritsoun [**9**] call it the "quasi-regular approximation".

This quasi-Gaussian approximation for FDT has been introduced by Leith in his seminal paper [**13**], and Bell [**2**], Dymnikov, Gritsoun and Branstator [**8, 9, 10**] have applied these formulas with some success to approximate the linear response to steady external forcing for a small number of leading low frequency patterns for prototype geophysical models. Intuitively, the approximation should be most successful for forced dissipative deterministic systems which are weakly divergent, $\| \operatorname{div}_{\vec{X}} \vec{F} \| \ll 1$, in important regions of phase space. In Section 2.8 we study the validity of the quasi-Gaussian approximation for FDT for forced dissipative systems defined by the Lorenz 96 model for "weather wave" packets for the atmosphere; we will find that in regimes of this model where $\| \operatorname{div}_{\vec{X}} \vec{F} \| \ll 1$ in a suitable fashion, the quasi-Gaussian approximation for FDT improves significantly compared with the ideal response operator.

In all the previously mentioned papers, the approximation in (2.92) and (2.93) has been tested for the special linear functionals, $\vec{A}(\vec{X}) = \vec{X}$ and for perturbed external forcing which is constant in time, i.e. the setup in (2.33) for external forcing. Given the success of this approximation under these circumstances, it is interesting to develop some mathematical theory supporting this. We do this next.

2.7.1. High accuracy of the quasi-Gaussian approximation for FDT for special functions, $A(\vec{X})$. We assume that the deterministic forced dissipative dynamics is given by a quadratically nonlinear system

$$\frac{d\vec{X}}{dt} = \mathcal{L}\vec{X} + \mathcal{B}(\vec{X}, \vec{X}) + \vec{F} \tag{2.94}$$

as occurs in the previous applications in atmospheric science [**2, 10**] and also in Section 2.8 below. The corresponding Fokker–Planck operator has quadratic coefficients in $\vec{X}$. We have the following

THEOREM 2.3. *Consider the forced dissipative quadratically nonlinear system in* (2.94) *and assume that it has a smooth invariant measure,* $p_{\text{eq}}(\vec{X})$. *Assume that the perturbation is defined by external forcing as in* (2.33). *Then*

- *For the special linear functionals,* $A(\vec{X}) = \vec{X}$, *the quasi-Gaussian approximation defined in* (2.92) *satisfies*

$$R^G_{\text{FDT}}(\tau) = R_{\text{FDT}}(\tau) + o(\tau^2), \tag{2.95}$$

- *For a quadratic functional,* $A(\vec{X}) = X_i X_j$, *the approximation in* (2.92) *satisfies*

$$R^G_{\text{FDT}}(\tau) = R_{\text{FDT}}(\tau) + o(\tau). \tag{2.96}$$

The content of the above theorem is that for the special functionals listed above, the approximation in (2.92) for the response operator is very good for short times. For the response to time-independent external forcing, which is defined in (2.67) through the time integral of $R(t)$, there is even better agreement so that $\mathcal{R}^G_{\text{FDT}}(t) = \mathcal{R}_{\text{FDT}}(t) + o(t^3)$ for the linear functionals, $A(\vec{X}) = \vec{X}$. Theorem 2.3 supplies some mathematical justification for the success of the quasi-Gaussian approximation for FDT. As the proof below shows, the requirement that $p_{\text{eq}}(\vec{X})$ is a smooth invariant measure for (2.94) probably is not necessary.

The formal proof of Theorem 2.3 begins with the formula for $R(\tau)$ in (2.21) with $\vec{L}_{\text{ext}} = -\nabla_{\vec{X}}$ so that

$$\begin{aligned}\vec{R}(\tau) &= -\int_{\mathcal{R}^N} A(\vec{X}) \exp[\tau L_{\text{FP}}][\nabla_{\vec{X}} p_{\text{eq}}]\, d\vec{X} \\ &= \int_{\mathcal{R}^N} p_{\text{eq}} \nabla_{\vec{X}} \big(\exp[\tau L^T_{\text{FP}}] A(\vec{X}) \big)\, d\vec{X}.\end{aligned} \tag{2.97}$$

Formally differentiating the second identity in (2.97) k times and evaluating at $\tau = 0$ yields

$$\frac{d^k}{d\tau^k} \vec{R}(\tau)\big|_{\tau=0} = \int_{\mathbb{R}^N} p_{\text{eq}}(\vec{X}) \nabla_{\vec{X}} \big[(L^T_{\text{FP}})^k A(\vec{X}) \big]\, d\vec{X}. \tag{2.98}$$

For the operator in (2.94), L^T_{FP} is a first order operator with coefficients which are second degree polynomials in $\vec{X}$; thus, if $A(\vec{X})$ is a polynomial of degree r, then

$$\nabla_{\vec{X}} (L^T_{\text{FP}})^k A(\vec{X}) \text{ is a polynomial of degree } k + r - 1. \tag{2.99}$$

On the other hand, by the construction in (2.92), for all quadratic polynomial functions, $Q(\vec{X})$,

$$\int Q(\vec{X}) p^G_{\text{eq}}(\vec{X})\, d\vec{X} = \int Q(\vec{X}) p_{\text{eq}}(\vec{X})\, d\vec{X}, \tag{2.100}$$

so (2.99) and (2.100) combine to finish the proof.

2.8. The Lorenz 96 model: A forced dissipative nonlinear system for testing FDT approximations

Here we discuss the Lorenz 96 model (L96) as a simple toy model with some statistical features of the weather wave packets of the atmosphere and utilize it to provide a comprehensive test of the validity of the quasi-Gaussian approximation to FDT for forced dissipative systems developed above in Section 2.7. We also carefully compare the behavior of L96 with the undamped unforced IL96 model which we studied in Section 2.6 both as regards its climatology and the FDT response to constant external forcing.

2.8.1. Geophysical properties of the Lorenz 96 model. As mentioned before, the L96 model possesses some general properties of geophysical models, namely energy-preserving advection, damping and forcing. The L96 model is given by

$$\frac{du_j}{dt} = (u_{j+1} - u_{j-2})u_{j-1} - u_j + F, \quad j = 0, \dots, J \tag{2.101}$$

with periodic boundary conditions, $u_0 = u_J$. The term $-u_j$ in (2.101) represents damping (with a unit time scale of 5 days) while F represents constant "solar forcing". The model in (2.101) is designed to mimic midlatitude weather and climate behavior, so periodic boundary conditions are appropriate. The unit spatial scale between discrete nodes is regarded as a non-dimensional midlatitude Rossby radius ≈ 800 km; the discrete system size is set to be $J = 40$ nodes so the discrete system with 40 modes corresponds to a midlatitude belt of roughly 30000 km.

In midlatitude weather systems, the main "weather waves", the Rossby waves have westward (toward negative x) phase velocity, but from our own anecdotal experience, weather systems collectively move eastward (toward positive x) with unstable behavior. The models in (2.101) have analogous behavior. To see this, note that $u_j = F$ defines a steady state solution of (2.101). We carry out this linear analysis next and confirm the above intuition.

2.8.1.1. *Linear stability of simple steady states.* With the steady solution, $u_j = F$, we write (2.101) in terms of perturbations around the steady state: taking

$$u_j = F + v_j,$$

the L96 model is written in new variables as

$$\frac{dv_j}{dt} = F(v_{j+1} - v_{j-2}) + (v_{j+1} - v_{j-2})v_{j-1} - v_j. \tag{2.102}$$

To check linear stability we linearize (2.102) and obtain

$$\frac{dv_j}{dt} = F(v_{j+1} - v_{j-2}) - v_j. \tag{2.103}$$

Proceeding with linear analysis of the Fourier modes, we introduce the discrete Fourier transform:

$$\hat{v}_k = \frac{1}{J}\sum_{j=0}^{J-1} v_j e^{-2\pi i k j/J}.$$

In terms of Fourier coefficients the equation (2.103) becomes

$$\frac{d\hat{v}_k}{dt} = a_k \hat{v}_k, \tag{2.104}$$

where

$$a_k = [(e^{2\pi ik/J} - e^{-4\pi ik/J})F - 1].$$

The real part of a_k is

$$\mathrm{Re}(a_k) = \left[\cos\left(\frac{2\pi k}{J}\right) - \cos\left(\frac{4\pi k}{J}\right)\right]F - 1,$$

and the linear instability occurs when $\mathrm{Re}(a_k) > 0$, or

$$\cos\left(\frac{2\pi k}{J}\right) - \cos\left(\frac{4\pi k}{J}\right) > \frac{1}{F}. \tag{2.105}$$

The next section offers detailed information about geophysical properties of the model based on these linearly unstable waves.

2.8.1.2. *Rossby waves.* As mentioned earlier, the L96 model has the analogues of Rossby waves for geophysical flows. We again write the linearized model (2.103):

$$\frac{dv_j}{dt} = F(v_{j+1} - v_{j-2}) - v_j.$$

Assuming v_j to be of the form

$$v_j = \sum_k A_k \exp\left(\frac{2\pi ikj}{J} - i\omega(k)t\right), \tag{2.106}$$

we substitute (2.106) into (2.103) and obtain the dispersion relation $\omega(k)$:

$$\omega(k) = i\left[\left(e^{2\pi ik/J} - e^{-4\pi ik/J}\right)F - 1\right]. \tag{2.107}$$

The imaginary part of $\omega(k)$ determines linear stability/instability of the wave, which has already been shown in the previous section. Denoting $\tilde{\omega}(k) \equiv \mathrm{Re}\left(\omega(k)\right)$, for $\tilde{\omega}(k)$ we write

$$\tilde{\omega}(k) = -\left[\sin\left(\frac{2\pi k}{J}\right) + \sin\left(\frac{4\pi k}{J}\right)\right]F. \tag{2.108}$$

The equation in (2.108) determines the phase and group velocities of the wave packet in (2.106). The phase velocity (velocity of a single wave) of (2.106) is defined as

$$V(k) = \frac{\tilde{\omega}(k)}{k} = -\left[\sin\left(\frac{2\pi k}{J}\right) + \sin\left(\frac{4\pi k}{J}\right)\right]\frac{F}{k}, \tag{2.109}$$

while the group velocity (velocity of a packet of waves) is defined as

$$V_{\mathrm{gr}}(k) = \frac{d\tilde{\omega}(k)}{dk} = -\frac{2\pi F}{J}\left[\cos\left(\frac{2\pi k}{J}\right) + \cos\left(\frac{4\pi k}{J}\right)\right]. \tag{2.110}$$

We denote the direction of positive V's as "eastward", and the direction of negative V's as "westward". The most interesting waves are those which have "westward" phase velocities like Rossby waves, and "eastward" group velocities like actual weather systems as discussed earlier in Chapter 1. From (2.109), we find that

- Waves with $(1/3 < k/J < 1/2)\bigcup(2/3 < k/J < 1)$ have "eastward" phase velocity,
- Waves with $(0 < k/J < 1/3)\bigcup(1/2 < k/J < 2/3)$ have "westward" phase velocity.

Looking at (2.110) we observe that

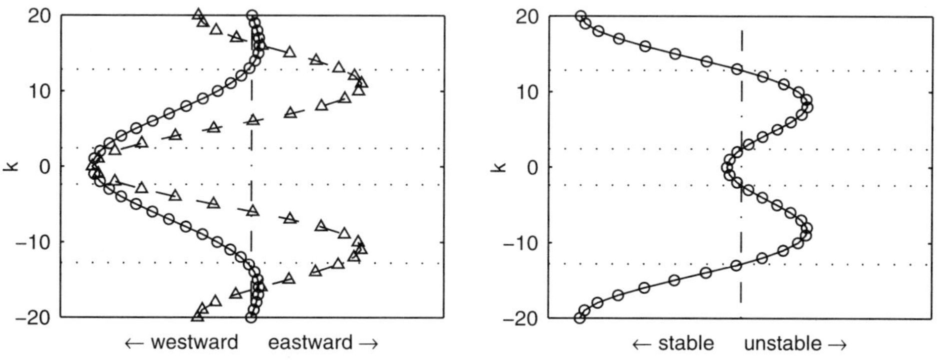

FIGURE 2.29. The phase/group velocities (left) and linearly stable/unstable wavenumbers (right) for $F = 5$ dynamical regime. Left plot: solid line with circles—phase velocity, dashed line with triangles—group velocity, vertical dash-dotted line marks zero velocity, horizontal dotted lines mark boundaries between stable and unstable regions. Right plot: solid line with circles is $[\cos(2\pi k/J) - \cos(4\pi k/J) - (1/F)]$, vertical dash-dotted line marks zero (neither stability nor instability), horizontal dotted lines mark boundaries between stable and unstable regions.

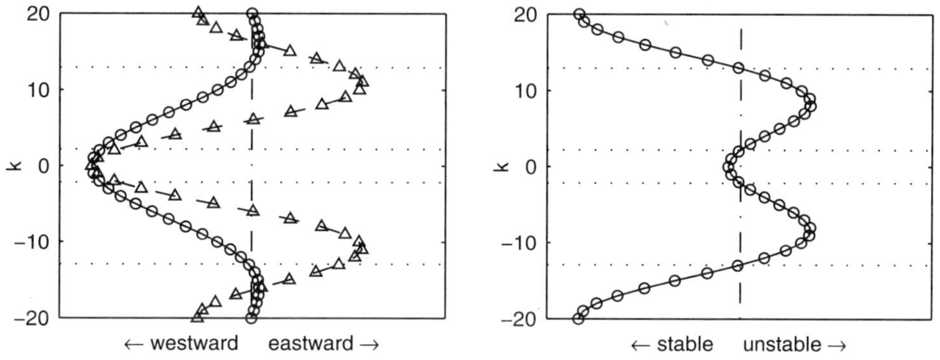

FIGURE 2.30. The phase/group velocities (left) and linearly stable/unstable wavenumbers (right) for $F = 6$ dynamical regime. Left plot: solid line with circles—phase velocity, dashed line with triangles—group velocity, vertical dash-dotted line marks zero velocity, horizontal dotted lines mark boundaries between stable and unstable regions. Right plot: solid line with circles is $[\cos(2\pi k/J) - \cos(4\pi k/J) - (1/F)]$, vertical dash-dotted line marks zero (neither stability nor instability), horizontal dotted lines mark boundaries between stable and unstable regions.

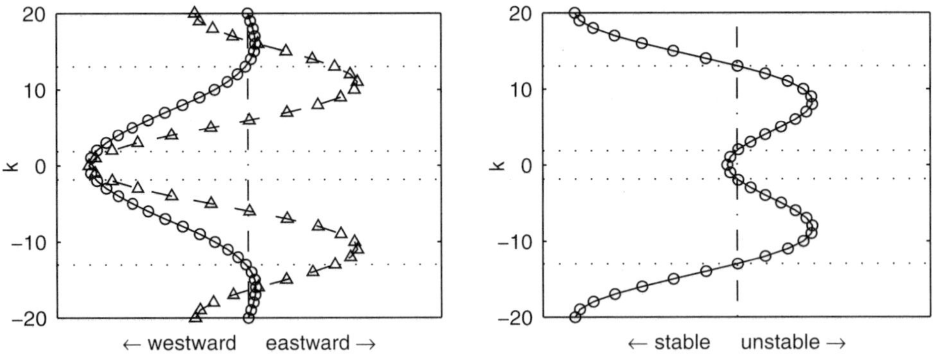

FIGURE 2.31. The phase/group velocities (left) and linearly stable/unstable wavenumbers (right) for $F = 8$ dynamical regime. Left plot: solid line with circles—phase velocity, dashed line with triangles—group velocity, vertical dash-dotted line marks zero velocity, horizontal dotted lines mark boundaries between stable and unstable regions. Right plot: solid line with circles is $[\cos(2\pi k/J) - \cos(4\pi k/J) - (1/F)]$, vertical dash-dotted line marks zero (neither stability nor instability), horizontal dotted lines mark boundaries between stable and unstable regions.

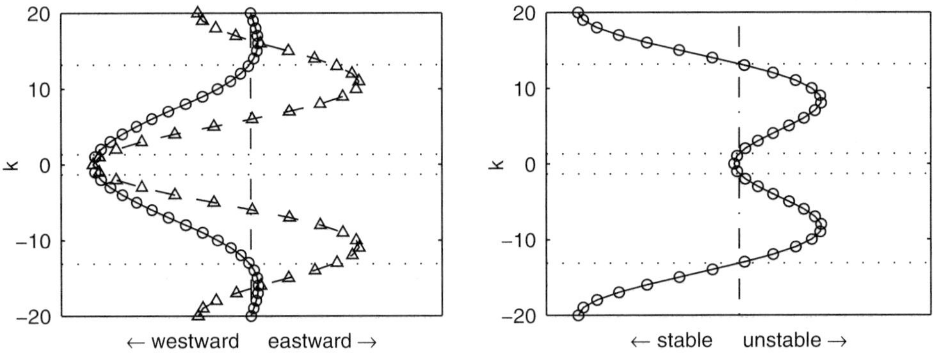

FIGURE 2.32. The phase/group velocities (left) and linearly stable/unstable wavenumbers (right) for $F = 16$ dynamical regime. Left plot: solid line with circles—phase velocity, dashed line with triangles—group velocity, vertical dash-dotted line marks zero velocity, horizontal dotted lines mark boundaries between stable and unstable regions. Right plot: solid line with circles is $[\cos(2\pi k/J) - \cos(4\pi k/J) - (1/F)]$, vertical dash-dotted line marks zero (neither stability nor instability), horizontal dotted lines mark boundaries between stable and unstable regions.

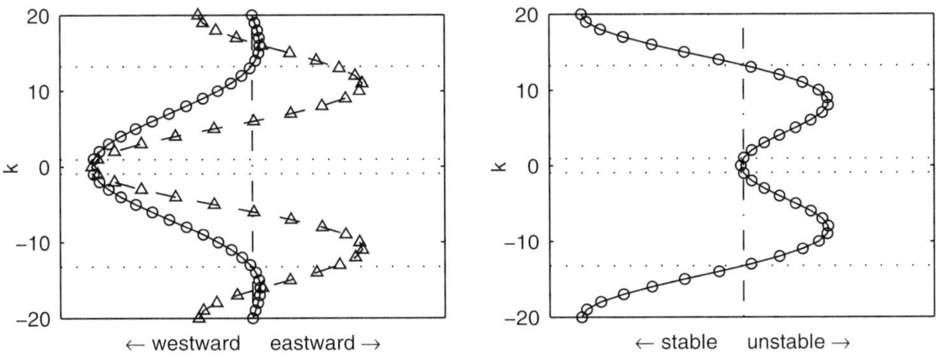

FIGURE 2.33. The phase/group velocities (left) and linearly stable/unstable wavenumbers (right) for $F = 32$ dynamical regime. Left plot: solid line with circles—phase velocity, dashed line with triangles—group velocity, vertical dash-dotted line marks zero velocity, horizontal dotted lines mark boundaries between stable and unstable regions. Right plot: solid line with circles is $[\cos(2\pi k/J) - \cos(4\pi k/J) - (1/F)]$, vertical dash-dotted line marks zero (neither stability nor instability), horizontal dotted lines mark boundaries between stable and unstable regions.

- Waves with $(0.1489 < k/J < 0.4096) \bigcup (0.5904 < k/J < 0.8511)$ have "eastward" group velocity,
- The rest of the waves has "westward" group velocity.

Combining the two results we obtain

$$\left(0.1489 < \frac{k}{J} < \frac{1}{3}\right) \bigcup \left(0.5904 < \frac{k}{J} < \frac{2}{3}\right). \tag{2.111}$$

The wavenumbers in (2.111) have "westward" phase and "eastward" group velocities.

Linearly unstable waves play an important role in the dynamics of the L96 model. The linear instability criterion is already given in (2.111). Figures 2.29–2.33 display phase and group velocities and stable and unstable wavenumbers for $J = 40$, and $F = 5$, 6, 8, 16 and 32. Note that wavenumbers of interest (which are linearly unstable and have phase and group velocities of opposite signs) generally lie between wavenumbers $k = 5$ and $k = 12$, and these correspond to the reasonable wave lengths of baroclinic instability in the atmosphere [**19**].

Below we call the solutions of (2.77) or (2.101) without damping and forcing the undamped unforced IL96 model and denote it by $F = 0$. In Figure 2.34 we demonstrate the propagation of typical solutions of the L96 model in time with constant forcing set to $F = 6$, $F = 8$ and $F = 0$ which emerge from numerical simulations at long times. These are typical solutions on the "attractor". One can observe that solutions for $F = 6$ and $F = 8$ definitely have patterns with opposite phase and group velocities, whereas the regime $F = 0$ is completely chaotic.

2.8.2. Statistical properties of the damped forced and undamped unforced L96 models. Here we compare and contrast the equilibrium statistical

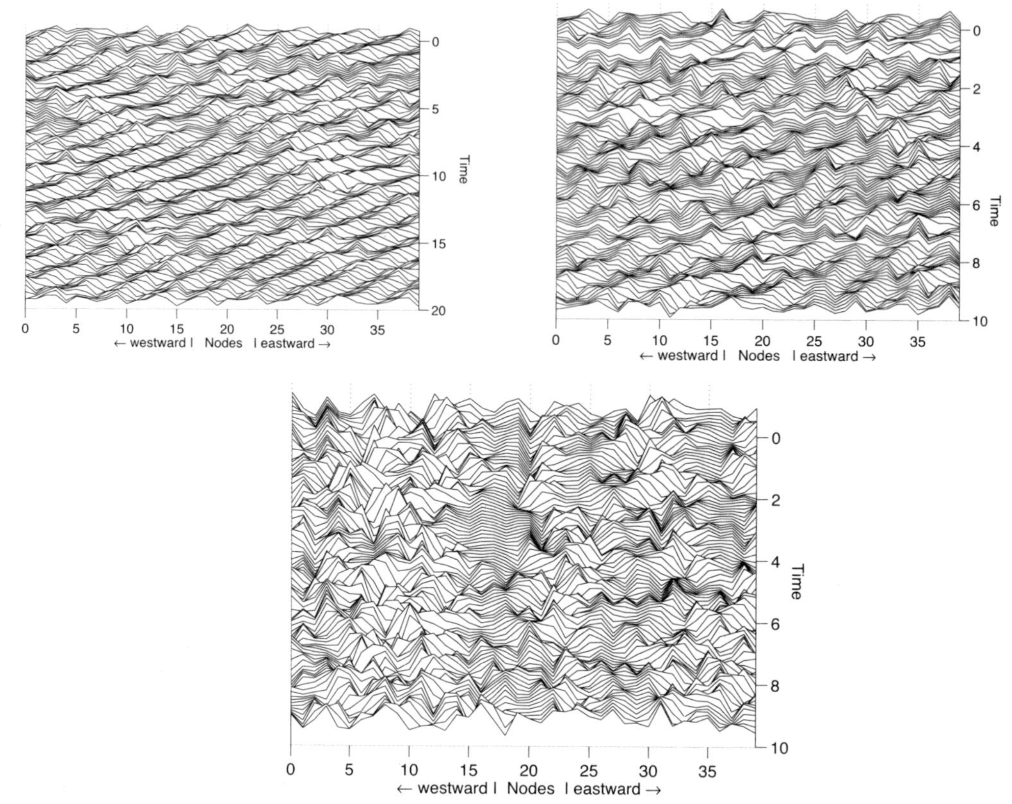

FIGURE 2.34. The snapshots of a numerical solution for the dynamical regimes $F = 6$ (upper left), $F = 8$ (upper right) and $F = 0$ (center).

predictions of Section 2.6 for IL96 with those for the L96 equation as we vary the forcing, F. It is interesting in many physical systems to understand how the statistical properties of a damped forced chaotic system like L96 are related to the predictions through equilibrium statistical mechanics of the corresponding undamped unforced equations, IL96. This comparison is what we will develop below through some simple theory and numerical experiments. We will find that many statistical features of IL96 are predicted and captured by L96 for moderately large values of F like $F = 8$, 16, 32. While more details are presented below, first we need to provide some intuition into these developments. First, as in Section 2.6, statistics are gathered for solutions of L96 from (2.101) or IL96 from long time averages. Thus, the statistics of a functional $f(\vec{u})$ of the variables, u_j, is given an empirical long time average

$$\overline{f(\vec{u})} = \frac{1}{T} \int_{T_0}^{T_0+T} f\big(\vec{u}(t)\big)\, dt \tag{2.112}$$

with T_0 being an initial startup time for the numerical solution to adjust to the attractor and T sufficiently large.

The first statistical difference between the undamped unforced IL96 model and the L96 model is that the simplest statistic, the ensemble mean, has a completely different character in these two models. In particular, $\bar{u}_j = 0$ for IL96, while $\bar{u}_j = \bar{u} \neq 0$ for L96, where for a given value of forcing F $\bar{u}$ is determined by a subtle balance of dissipation, forcing and nonlinearity. We note that the L96 model in (2.101) is translation invariant so the statistical ensemble mean at any given coordinate value is necessarily the same for each j and denoted by $\bar{u}$ with $\bar{u}$ being a nonlinear function of F. This statistical mean value $\bar{u}$ is called the climatology below by analogy with realistic geophysical systems. In Table 2.16 below we tabulate $\bar{u}$ and we see that $\bar{u}$ from the simulations is nearly zero, i.e. 10^{-3} for IL96, while $\bar{u}$ is clearly a nonlinear function of F for $F = 5, 6, 8, 16, 32$ which does not coincide with the trivial steady state $u_j = F$ utilized in the discussion in Section 2.8.1. The energy of this mean state, $E(\bar{u})$, is given by

$$E(\bar{u}) = \frac{J}{2}\bar{u}^2. \tag{2.113}$$

The second statistical quantity of interest for L96 is the strength of fluctuations about the given climatology; this is measured by the average variance in energy fluctuations about the mean,

$$E_p = \frac{1}{2}\sum_{j=0}^{J-1} \overline{(u_j - \bar{u})^2}. \tag{2.114}$$

Once again for the undamped unforced IL96, there is the precise equilibrium statistical prediction for (2.114), $E_p = E$, the predicted energy of fluctuations, already verified in Section 2.6; on the other hand, $E_p(F)$ is a nonlinear function of the forcing F which involves subtle balances of forcing, dissipation and nonlinearity with the values of $E_p(F)$ plotted in Table 2.16. It is clear from Table 2.16 that the energy of the fluctuations dominates the energy of the mean as F increases. We contend that as F increases the statistics of the L96 model are in many respects converging to those of IL96 established in Section 2.6.3. To facilitate such a comparison it is natural to rewrite the L96 model in rescaled variables using $\bar{u}$ and E_p so that comparisons can be made with equilibrium statistical mechanics predictions for a fixed energy.

2.8.2.1. *Rescaling the damped forced* L96 *model.* The simple idea of the rescaling $u_j \to \tilde{u}_j$, $t \to \tilde{t}$ is to set the mean state to zero in new coordinates, and energy of perturbations to 1 while retaining the same form of the nonlinear term in the right-hand side of the L96 model in (2.101). The corresponding coordinate-time rescaling is given by

$$u_j = E_p^{1/2}\tilde{u}_j + \bar{u}, \quad t = E_p^{-1/2}\tilde{t}. \tag{2.115}$$

The L96 model in (2.101), written through the coordinates $\tilde{u}_j$ and $\tilde{t}$, becomes

$$\frac{d\tilde{u}_j}{d\tilde{t}} = (\tilde{u}_{j+1} - \tilde{u}_{j-2})\tilde{u}_{j-1} + E_p^{-1/2}[(\tilde{u}_{j+1} - \tilde{u}_{j-2})\bar{u} - \tilde{u}_j] + E_p^{-1}(F - \bar{u}). \tag{2.116}$$

Obviously, if the forcing F increases, both the mean state $\bar{u}$ and the energy of perturbations E_p will also increase as shown in Table 2.15. Thus, if the limits

$$E_p^{-1/2}\bar{u} \to 0 \quad \text{and} \quad E_p^{-1}(F - \bar{u}) \to 0 \quad \text{as} \quad F \to \infty \tag{2.117}$$

are valid, then the rescaled dynamics in (2.116) should approach the dynamics of the L96 model with no damping and forcing with energy $E_p = 1$.

TABLE 2.15. Bands of linearly stable and unstable waves for different values of F and $\bar{u}$, $J = 40$.

F	$\bar{u}$	k_{begin}	k_{end}
5	1.63668	5	11
6	2.0109	4	12
8	2.34299	4	12
16	3.08713	3	12
32	3.89012	3	12
∞	∞	0	13

2.8.2.2. *Linear stability of the mean state.* The motion of the rescaled system in (2.116) is centered, by construction, around its statistical mean state $\bar{u}$. Thus, it is important to check whether its solution is linearly stable or unstable for small $\tilde{u}_j$. Following (2.103), the rescaled linearized system is written in Fourier space as

$$\dot{\hat{u}}_k = \omega(k)\hat{u}_k + E_p^{-1}(F - \bar{u}), \quad \omega(k) = E_p^{-1/2}[\bar{u}(e^{2\pi i k/J} - e^{-4\pi i k/J}) - 1] \tag{2.118}$$

where J denotes the number of gridpoints in the model. The linear stability criterion is

$$\begin{gathered} \operatorname{Re}\omega(k) = E_p^{-1/2}\left[\bar{u}\left(\cos(2\pi k/J) - \cos(4\pi k/J)\right) - 1\right], \\ \operatorname{Re}\omega(k) < 0 \quad \text{stable}, \\ \operatorname{Re}\omega(k) > 0 \quad \text{unstable}. \end{gathered} \tag{2.119}$$

Solving the equation $\operatorname{Re}\omega(k) = 0$, we find out that the band of unstable waves emerges when

$$\bar{u} > 8/9.$$

It begins at

$$k_{\text{begin}} = \frac{J}{2\pi}\arccos\left(\frac{1 + 3\sqrt{1 - 8/9\bar{u}^{-1}}}{4}\right), \tag{2.120}$$

and ends at

$$k_{\text{end}} = \frac{J}{2\pi}\arccos\left(\frac{1 - 3\sqrt{1 - 8/9\bar{u}^{-1}}}{4}\right). \tag{2.121}$$

In particular, in the limiting case with $F \to \infty$ and, therefore, $\bar{u} \to \infty$

$$k_{\text{begin}} = 0, \quad k_{\text{end}} = \frac{J}{3}. \tag{2.122}$$

The boundaries of the band of stable and unstable waves are shown in the Table 2.15 for different F and $\bar{u}$ for the number of gridpoints $J = 40$. Note that even as $F \to \infty$, the wavenumbers beyond $k = 13$ remain linearly stable in the formal limit.

2.8.2.3. *The bulk behavior of the rescaled problem.* Numerical solutions of both the damped forced L96 model in (2.101) and the IL96 model without forcing and damping in (2.77) are found by numerical time integration via the fourth-order Adams–Bashforth multistep method with the time step $\Delta t = 1/64$. All statistical experiments are carried out by a long-time integration of the two models on a time averaging window of $T = 100000$ units with the initial spin-up time of $T_0 = 1000$ time units. The computations are performed for the forced damped regimes $F = 5$,

TABLE 2.16. The values of energy of the mean $E(\bar{u})$ and the energy of perturbations E_p for different dynamical regimes.

F	$\bar{u}$	$E(\bar{u})$	E_p	$\bar{u}/\sqrt{E_p}$
0	0.004	0.0003	9.98634	$0.00113 \cdot 10^{-2}$
5	1.63668	53.5745	110.091	$0.156 \cdot 10^{-2}$
6	2.0109	80.8742	160.425	$0.1588 \cdot 10^{-2}$
8	2.34299	109.792	265.028	$0.1439 \cdot 10^{-2}$
16	3.08713	190.607	797.126	$0.1093 \cdot 10^{-2}$
32	3.89012	302.661	2186.68	$0.08319 \cdot 10^{-2}$

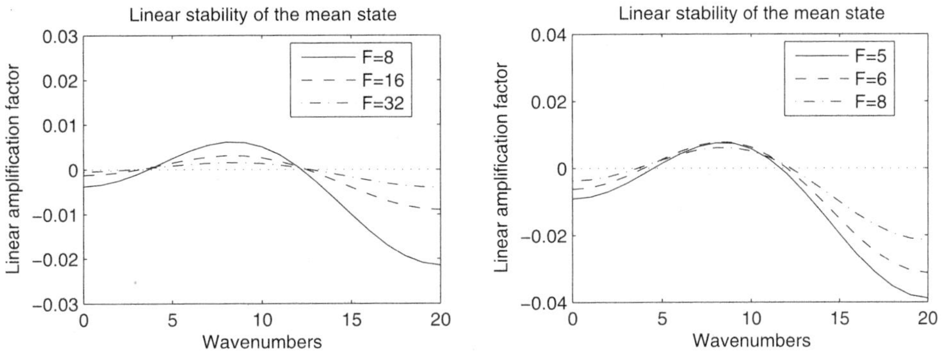

FIGURE 2.35. The linear amplification factor $\mathrm{Re}\,\omega(k)$ of the linearization about the mean state $\bar{u}$ for the regimes $F = 5$, 6, 8, 16, 32.

6, 8, 16, 32 and the regime without forcing and damping for IL96 (sometimes denoted below as the $F = 0$ regime).

The decay of the quantity $E_p^{-1/2}\bar{u}$ in (2.117) is equivalent to the decay of the ratio between the energy of the mean state $E(\bar{u})$ and the mean energy of perturbations E_p, and we choose to show the ratio of energies instead. Magnitudes of forcing F and corresponding values of $E(\bar{u})$ and E_p are shown in the Table 2.16. It can be seen that E_p grows faster than $E(\bar{u})$ as F increases. One can also observe that despite the fact that the band of linearly unstable waves grows as energy increases, the magnitude of the linear amplification factor $\mathrm{Re}\,\omega(k)$ from (2.118) decays to zero at the same time, which is shown in Figure 2.35. This means that as $F \to \infty$, the linearly unstable waves become almost neutrally stable.

In Figure 2.36 we demonstrate the experimental dependence of energy and mean state ratios from (2.117) on the forcing F. The left plot of Figure 2.36 demonstrates how the ratio $E(\bar{u})/E_p$ depends on the forcing F (reciprocal scale $1/F$ is shown along the horizontal axis). All three results for the dynamical regimes $F = 8$, $F = 16$ and $F = 32$ lie approximately on a straight line which goes through the origin. Though it can not be established convincingly with only three data points, still the assumption that the ratio of energy of the mean state and mean energy of perturbations goes to zero as the forcing F increases to infinity seems to be valid. The right plot of Figure 2.36, which demonstrates the dependence of the

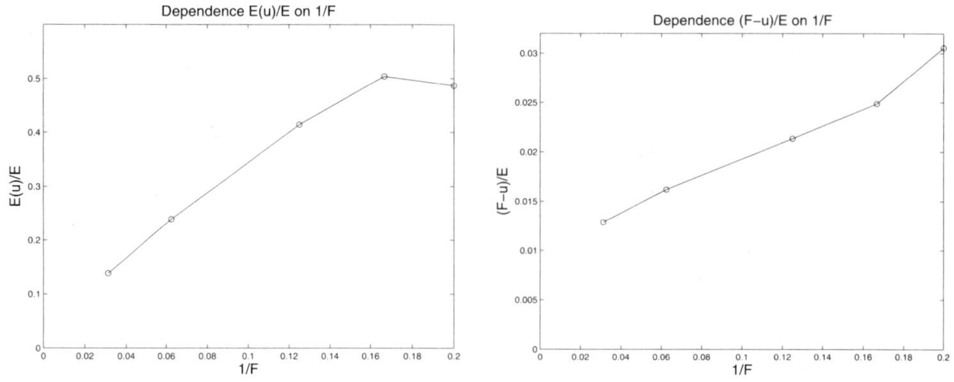

FIGURE 2.36. Left: the dependence of $E(\bar{u})/E_p$ on $1/F$. Right: the dependence of $(F-\bar{u})/E_p$ on $1/F$.

ratio $(F-\bar{u})/E_p$ on the forcing F, does not, however, show the convergence to zero as F increases, but tends toward a finite value in this range of F.

2.8.2.4. *The climatology of different forcing regimes in rescaled coordinates.* Earlier we made an assumption that with increasing forcing F in the damped forced L96 model, the rescaled dynamics in (2.116) should approach the dynamics of the undamped unforced IL96 model. If that assumption holds, one should be able to observe the statistical similarities in the climatology of the L96 model in the rescaled space-time coordinates. The rescaled variance spectrum of the L96 model is shown in Figure 2.37 for the regimes $F = 0$, 5, 6, 8, 16 and 32. The variance spectrum is defined as a set of time averaged variances for the Fourier coefficients

$$\mathrm{Var}(\hat{u}_k) = \overline{|\hat{u}_k(t) - \bar{\hat{u}}_k|^2}, \tag{2.123}$$

where the overbar denotes the time average, such that $\bar{\hat{u}}_k$ is the time average of $\hat{u}_k(t)$. In practice the time averaging window T is finite but large (10^5 time units in our simulations). As already established in Section 2.6, for $F = 0$ the spectrum is flat as expected, and strongly confirms the equipartition of energy for IL96. The rescaled spectra for $F \neq 0$ are nonuniform, with a hump where the band of linearly unstable wave modes is located, however one can see that in the part of the spectrum for the wavenumbers $k = 0, \ldots, 10$ there is a systematic trend: the spectrum for $F \neq 0$ in that part tends to approach the undamped unforced flat spectrum as F increases. The same trend, however, does not hold for the range of wavenumbers $k = 13, \ldots, 20$, where the spectrum $F \neq 0$ does not systematically approach the inviscid case. The possible reason could be that these wavenumbers remain linearly stable at the climatological mean state even when $F \to \infty$, according to Table 2.15 and our earlier discussion.

The time autocorrelation functions are shown in Figures 2.38 and 2.39 for the variety of linearly stable and unstable wavenumbers $k = 0, 3, 8, 12$ for the dynamical regimes $F = 0$, 5, 6, 8, 16 and 32. Recall the general formula for the autocorrelation function of a variable $a(t)$ is written as

$$\mathrm{Corr}(\tau) = \frac{\overline{(a(t)-\bar{a})(a(t+\tau)-\bar{a})}}{\overline{(a(t)-\bar{a})^2}}, \tag{2.124}$$

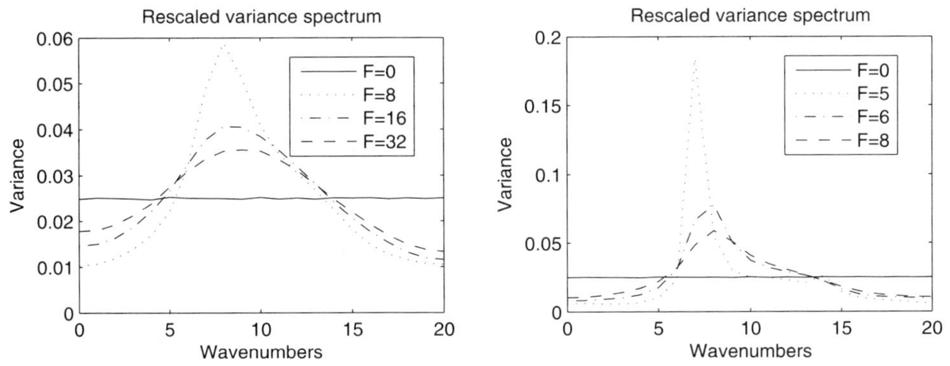

FIGURE 2.37. The variance spectrum for the L96 model in different dynamical regimes. Left plot: solid line—$F = 0$, dotted line—$F = 8$, dot-dashed line – $F = 16$, dashed line—$F = 32$. Right plot: solid line—$F = 0$, dotted line – $F = 5$, dot-dashed line—$F = 6$, dashed line—$F = 8$. Note the much smaller scale for the y-axis on the left.

where overbar denotes the time average, such that $\bar{a}$ is the time average of $a(t)$. In practice the time averaging window T is finite but large (10^5 time units in our simulations). The autocorrelation function in (2.124) measures mixing in a dynamical system, and in some sense can be regarded as an indicator of how fast a variable "forgets" about its behavior in the past. Each plot on Figures 2.38 and 2.39 reveals the following fact: first there is a stage until approximately time $t = 5 - 10$ (depending on the wavenumber shown) where there are no significant differences in the time autocorrelations between different wave numbers with various stability characteristics. After that time, however, the time autocorrelation functions for different dynamical regimes diverge from each other. The remarkable observation here is that while the time correlations for a linearly unstable wavenumber like $k = 8$ and high values of F like $F = 16$ and $F = 32$ are close to the non-oscillatory correlation for $F = 0$ which decays fast (Figure 2.38), the correlation functions for $F = 5, 6, 8$ are highly oscillatory and exhibit rather weak decay in time (Figure 2.39). There is a systematic trend: the larger the forcing is, the closer the corresponding correlation function is to the case with no forcing and no damping. The observed results on Figures 2.38 and 2.39 generally confirm the assumption that for large values of the forcing F the rescaled dynamics in (2.116) roughly approaches that of the undamped unforced IL96 model.

As discussed earlier for the TBH model, the correlation time is defined as a typical time of the decay of the autocorrelation function. Usually it is given through the time integral of the autocorrelation function. Here we look at two measures of the correlation time: one is the integral of the autocorrelation function

$$T = \int_0^\infty \operatorname{Corr}(\tau)\, d\tau, \tag{2.125}$$

and the other is the integral of the absolute value of the autocorrelation function

$$T_{\text{abs}} = \int_0^\infty |\operatorname{Corr}(\tau)|\, d\tau. \tag{2.126}$$

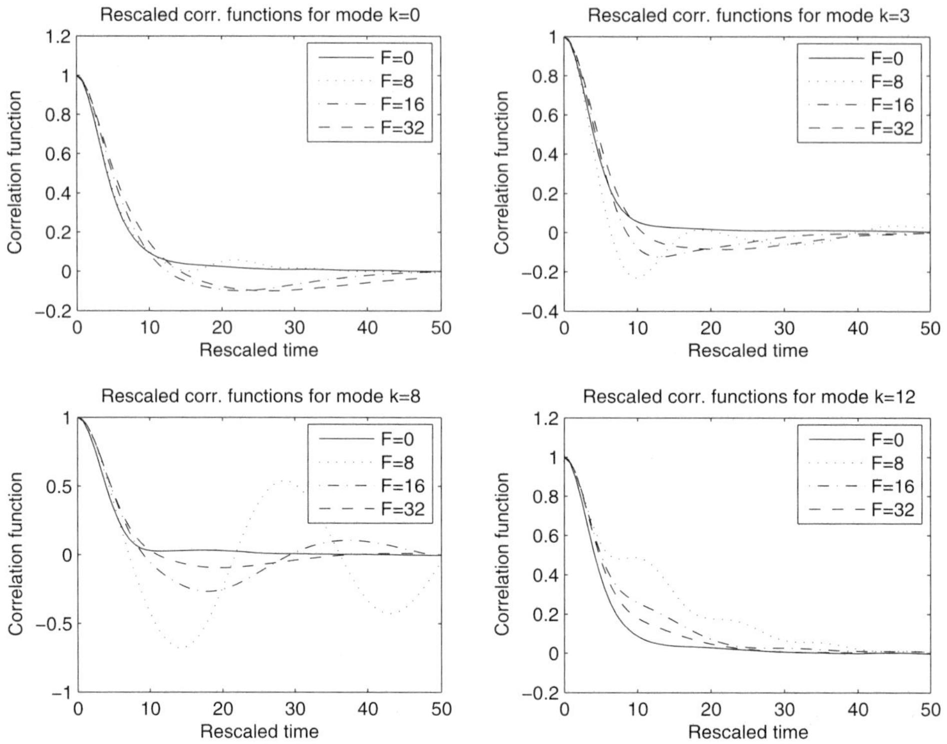

FIGURE 2.38. The time autocorrelation functions for the wavenumbers $k = 0, 3, 8, 12$, L96 model in different dynamical regimes. Solid line—$F = 0$, dotted line—$F = 8$, dot-dashed line—$F = 16$, dashed line—$F = 32$.

The first measure allows for oscillatory cancellation in the correlations while the second does not. The two types of correlation time spectra are presented in Figure 2.40: the correlation times on the left plots of Figure 2.40 functions, whereas the correlation times on the right plots of Figure 2.40 are obtained by integrating the absolute value of the corresponding correlation function. The correlation time spectra for the $F \neq 0$ regimes using the measure in (2.125) tend approach the unforced undamped case as F increases. The highly oscillatory correlation functions for linearly unstable wavenumbers in the regimes $F = 5$, 6, 8 produce completely different correlations T and T_{abs}: while T are even smaller that those for the IL96 model due to cancellation of positive and negative parts of correlation functions, T_{abs} are huge, differing almost by an order of magnitude. Also note that Figure 2.40 for $F = 0$, i.e. IL96, shows that all Fourier modes decay at the same rate in this conservative model, in contrast to what we established in Section 2.6.1 for the TBH model.

2.8.3. The validity of quasi-Gaussian FDT for the L96 model. Similarly to the conservative TBH, K-Z and IL96 models in Section 2.6, here we present systematic numerical experiments testing the applicability and validity of the quasi-Gaussian FDT from Section 2.7 with regard to the damped forced L96 model in

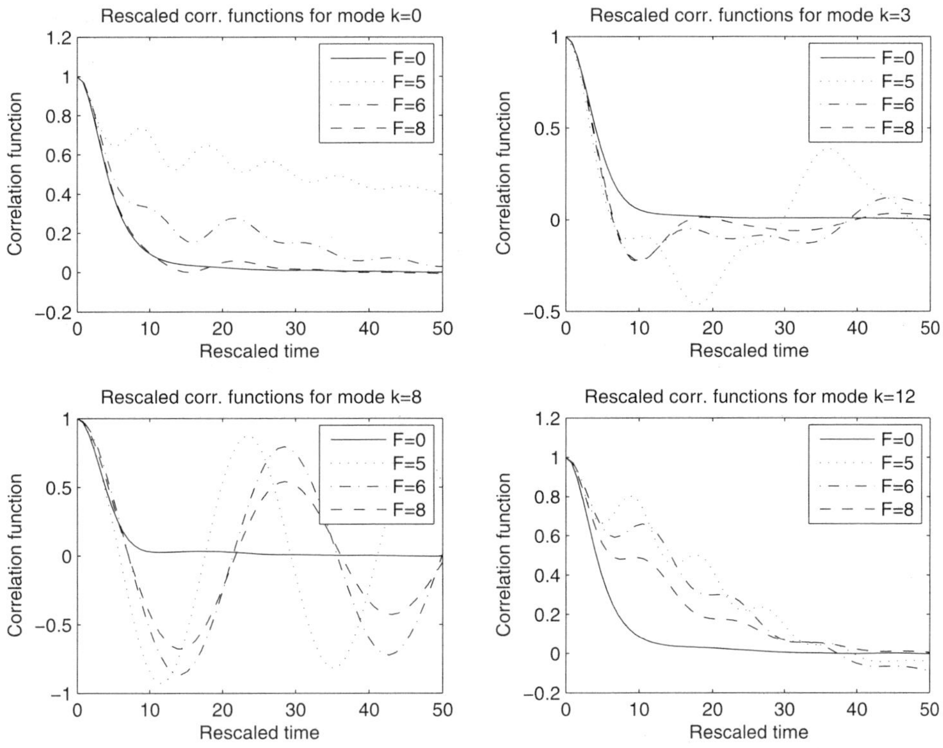

FIGURE 2.39. The time autocorrelation functions for the wavenumbers $k = 0, 3, 8, 12$, L96 model in different dynamical regimes. Solid line—$F = 0$, dotted line—$F = 5$, dot-dashed line—$F = 6$, dashed line—$F = 8$.

(2.116) in rescaled coordinates, which is discussed earlier in this section. We adopt the same style of presentation as in Section 2.6, i.e. first the evolution of singular values is presented for both the ideal response measurements and their FDT predictions, followed by more detailed discussion of the FDT behavior and validity at selected individual time slices. While in Section 2.6 we had three separate subsections describing the FDT validation for TBH, K-Z and IL96 models, respectively, here we present the results of numerical experiments for the single forced-dissipative L96 model. The L96 model is nonetheless studied in various dynamical regimes with different constant forcing F and compared to the IL96 model in the rescaled limit as $F \to \infty$, so the overall bulk of presented observations is comparable to that in Section 2.6. As well as making the direct comparison of FDT and ideal responses for different values of F, we also discuss the "convergence" of results for forced-dissipative regimes to IL96 in the limit $F \to \infty$ for both FDT prediction and measured ideal responses. The infinite time linear response results for the L96 model are presented in the end of the section. Note that the results for IL96 reproduced for comparison here are denoted by $F = 0$ in all graphs.

2.8.3.1. *Time evolution of singular values.* In Figure 2.41 we compare and contrast the time evolution of the largest singular value of the ideal and FDT results

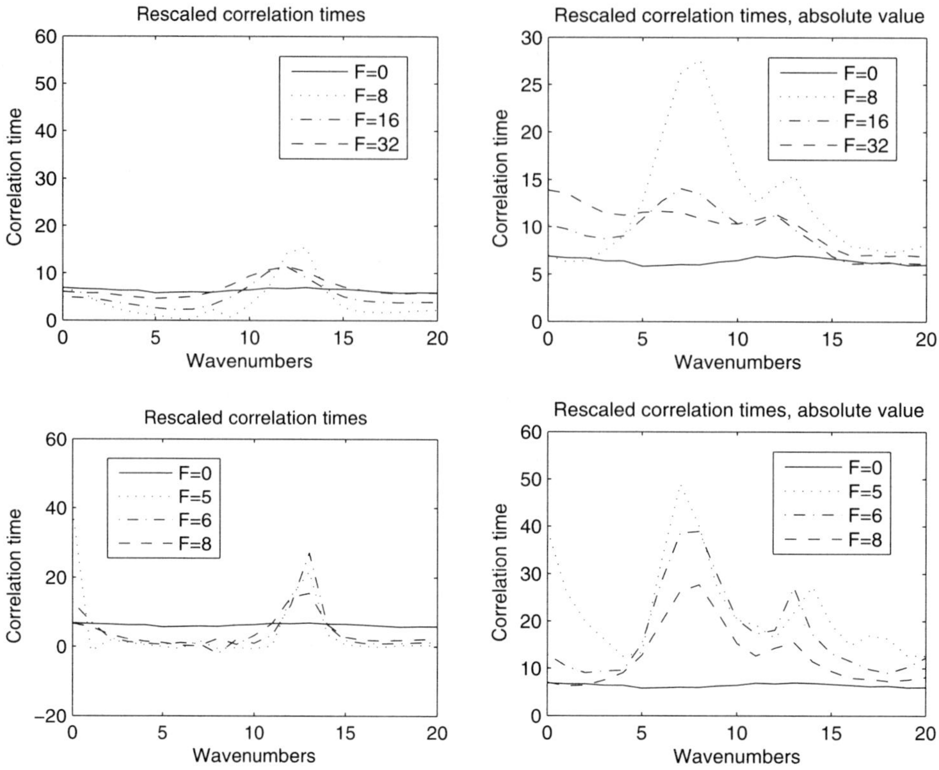

FIGURE 2.40. The spectra of correlation times (integrals of correlation functions) for the L96 model in different dynamical regimes. Left plots: the integrals are taken with the time correlation functions directly. Right plots: the integrals are taken with the absolute values of the correlation functions. Upper plots: solid line —$F = 0$, dotted line—$F = 8$, dot-dashed line—$F = 16$, dashed line —$F = 32$. Lower plots: solid line—$F = 0$, dotted line—$F = 5$, dot-dashed line—$F = 6$, dashed line—$F = 8$.

for different values of F in rescaled coordinates. Each plot of Figure 2.41 contains two curves, solid and dashed, representing the ideal response measurement and its quasi-Gaussian FDT prediction, respectively. Observe that the intervals of FDT validity are different depending on the value of the constant forcing F. For $F = 0$ regime the FDT is valid for the whole time scale displayed, until time $T = 30$. For $F = 5$ the time interval of FDT validity is until approximately $T = 5$, after which it diverges from the ideal result. The FDT for $F = 6$ regime start diverging at time about $T = 10$. Surprisingly, the FDT for $F = 8$ follows the actual result until time $T = 20$, which is also true for the regimes $F = 16$ and $F = 32$. Overall, the validity of quasi-Gaussian FDT approximation seems to improve systematically with increasing values of F. In Figure 2.42 we separately overlay the time series of the largest singular value of the response in different regimes for both the FDT prediction and ideal measurements to check the ansatz in (2.117). It turns out that the

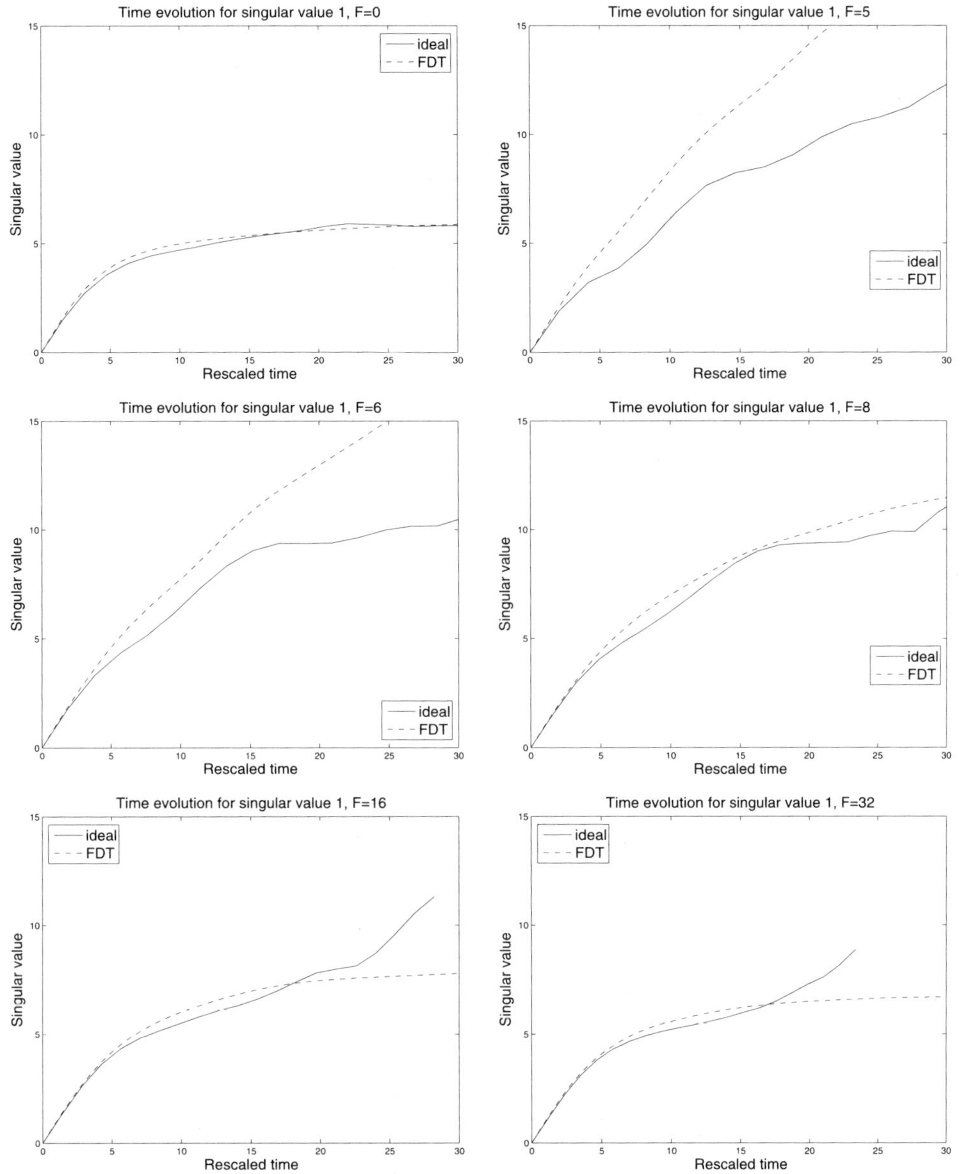

FIGURE 2.41. The time evolution for singular value 1 for the L96 model in rescaled coordinates, regimes $F = 0, 5, 6, 8, 16, 32$.

scaling argument from (2.117) is strongly confirmed by the results in Figure 2.42, that is, the time series for the forced-dissipative L96 model systematically approach those for the IL96 model. We do not show the rest of the singular values here due to their similar behavior in time. The reason for such behavior is the absence of time autocorrelation scaling with wavenumbers (shown in Figure 2.40), so that all the Fourier modes decorrelate simultaneously. Below we study the linear response

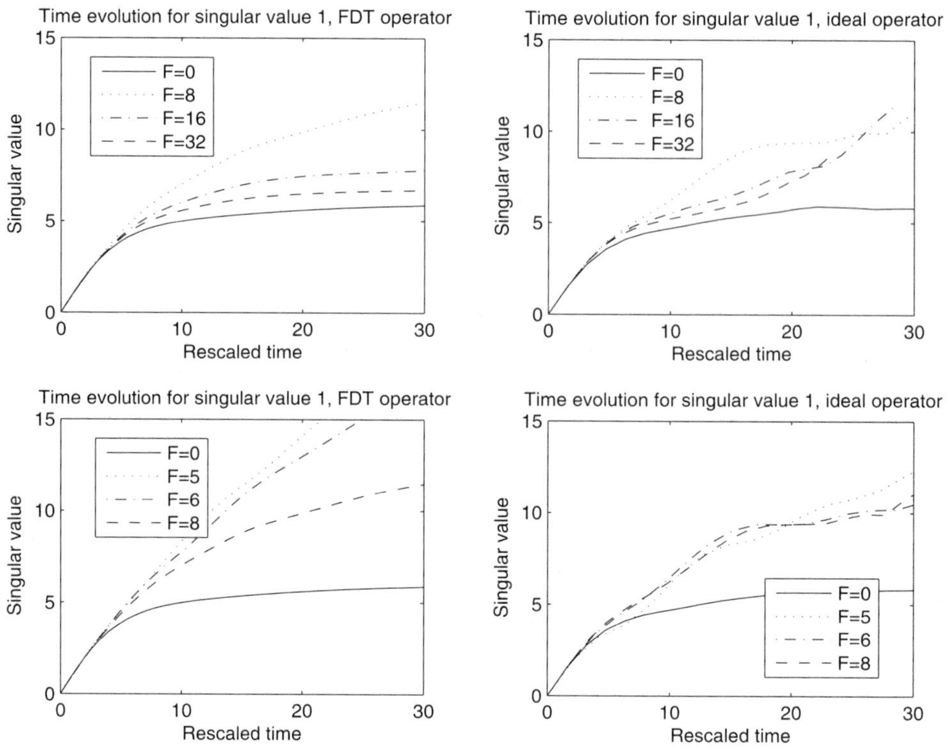

FIGURE 2.42. The evolution of singular value 1 for the L96 model in rescaled coordinates. Left plot—FDT approximation, right plot—ideal operator. Dynamical regimes $F = 0, 5, 6, 8, 16, 32$.

in detail for the time $T = 5$ (that is where the FDT singular values still match the ideal response), $T = 10$ and the infinite time.

2.8.3.2. *Linear response operators for the time* $T = 5$. Here we show the properties of the linear response operators at the time $T = 5$. In Figure 2.43 we check the validity of the FDT approximations for the regimes $F = 0, 5, 6, 8, 16, 32$ by showing the real parts of the main diagonals of both FDT and ideal linear response operators in the Fourier space. As we can see, the FDT approximation for the regime $F = 0$ is almost perfect, however the forced damped regimes develop some discrepancies. In general the FDT consistently overshoots the ideal response. Observe that the spectral part of the linear response operator which is never captured for the damped forced regimes is roughly the range of linearly stable wavenumbers $13 \leq k \leq 20$ and $o \leq k \leq 3$ defined by (2.122) for $J = 40$. On the other hand, the part of the spectrum which is captured better than the rest is the band of linearly unstable wavenumbers $5 \leq k \leq 12$ (see, for example, the plot for $F = 8$ in Figure 2.43). This remarkable observation has important consequences. On one hand, the precision of the quasi-Gaussian FDT for the forced damped regimes in L96 model at the time $T = 5$ is not that great (if compared, for example, to the $F = 0$ regime, shown in the upper-left corner of Figure 2.43). On the other hand, recall from the previous section that the band of unstable wavenumbers $5 \leq k \leq 12$ is actually

TABLE 2.17. The errors and physical correlations for the regimes $F = 0, 5, 6, 8, 16, 32$, L96 model, at time $T = 5$. The results are displayed for real and imaginary parts of linear response operators in Fourier space, the translationally averaged linear response operators in physical space, and the spectra of singular values.

$F = 0$, time $T = 5$		
	L^2 error	$C(g, h)$
Fourier diag.	0.04358	0.9999
Phys. av.	0.0422	0.9999

$F = 5$, time $T = 5$		
	L^2 error	$C(g, h)$
Fourier diag.	0.2031	0.9883
Phys. av.	0.4283	0.9148

$F = 6$, time $T = 5$		
	L^2 error	$C(g, h)$
Fourier diag.	0.3622	0.9883
Phys. av.	0.4046	0.9718

$F = 8$, time $T = 5$		
	L^2 error	$C(g, h)$
Fourier diag.	0.2245	0.9834
Phys. av.	0.2577	0.9707

$F = 16$, time $T = 5$		
	L^2 error	$C(g, h)$
Fourier diag.	0.245	0.9926
Phys. av.	0.2395	0.9924

$F = 32$, time $T = 5$		
	L^2 error	$C(g, h)$
Fourier diag.	0.1424	0.9927
Phys. av.	0.14	0.992

the most important physical part of the L96 model as it mimics real-world Rossby waves with oppositely directed phase and group velocities, and this is the part of the response spectrum which is captured better than the others. Figure 2.44 shows the translationally averaged linear response operators in physical space for the values of forcing $F = 5, 6, 8, 16, 32$ as well as the undamped unforced regime with $F = 0$. One can see that the response for the undamped unforced regime $F = 0$ is essentially a δ-function at the perturbation gridpoint with small oscillations elsewhere, whereas responses for $F \neq 0$ show some asymmetric off-diagonal response. The undamped unforced regime $F = 0$ shows perfect matching of ideal and FDT operators, however the forced damped regimes have some discrepancies between FDT and ideal response operators. It can be clearly seen that with increasing F, the FDT approximation improves significantly. Here we can also see that the FDT generally overshoots the ideal results. Again, observe that the large-magnitude bulk features of the ideal response operator are captured rather well for all regimes with minor discrepancies with small magnitudes. In Table 2.17 we display L^2 relative error charts defined in (2.84) and physical correlations from (2.85). According to the Table 2.17 the FDT approximation improves a lot when F increases. Note that although the L^2 errors are large in some cases with small F, the correlations $C(g, h)$ are very close to 1 in all cases, which means that the FDT is highly collinear with the ideal response, while consistently overshooting it, as we saw earlier.

2.8.3.3. *Linear response operators for the time $T = 20$.* Here we show the properties of the linear response operators at the time $T = 20$. In Figure 2.45 we check the validity of the FDT approximations for the regimes $F = 0, 5, 6, 8, 16, 32$ by showing the real parts of the main diagonals of both FDT and ideal linear response operators in the Fourier space. As we can see, the FDT approximation for the regime $F = 0$ is almost perfect, however the forced damped regimes develop some

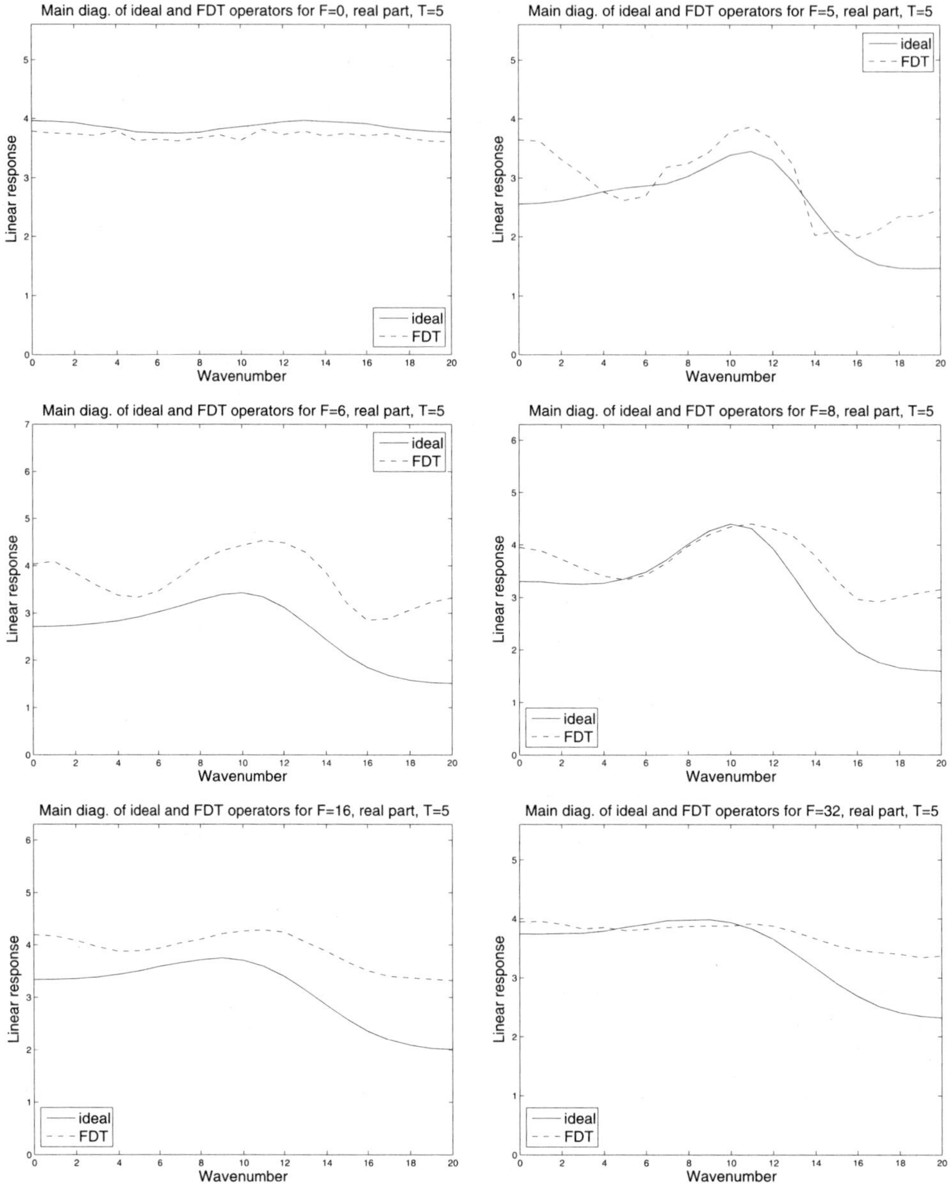

FIGURE 2.43. The real part of the main diagonal of the linear response operator in Fourier space. Comparison between FDT and ideal response. L96 model, dynamical regimes $F = 0$, 5, 6, 8, 16, 32. Time $T = 5$.

discrepancies. Again, as for the time $T = 5$, the FDT consistently overshoots the ideal response. However, the quasi-Gaussian FDT behavior here is qualitatively different here. Observe that the in general the FDT recovers the shape of the main diagonal of the ideal response operator, however it seems to be shifted towards smaller scales by one or two wavenumbers. This effect can be seen especially well

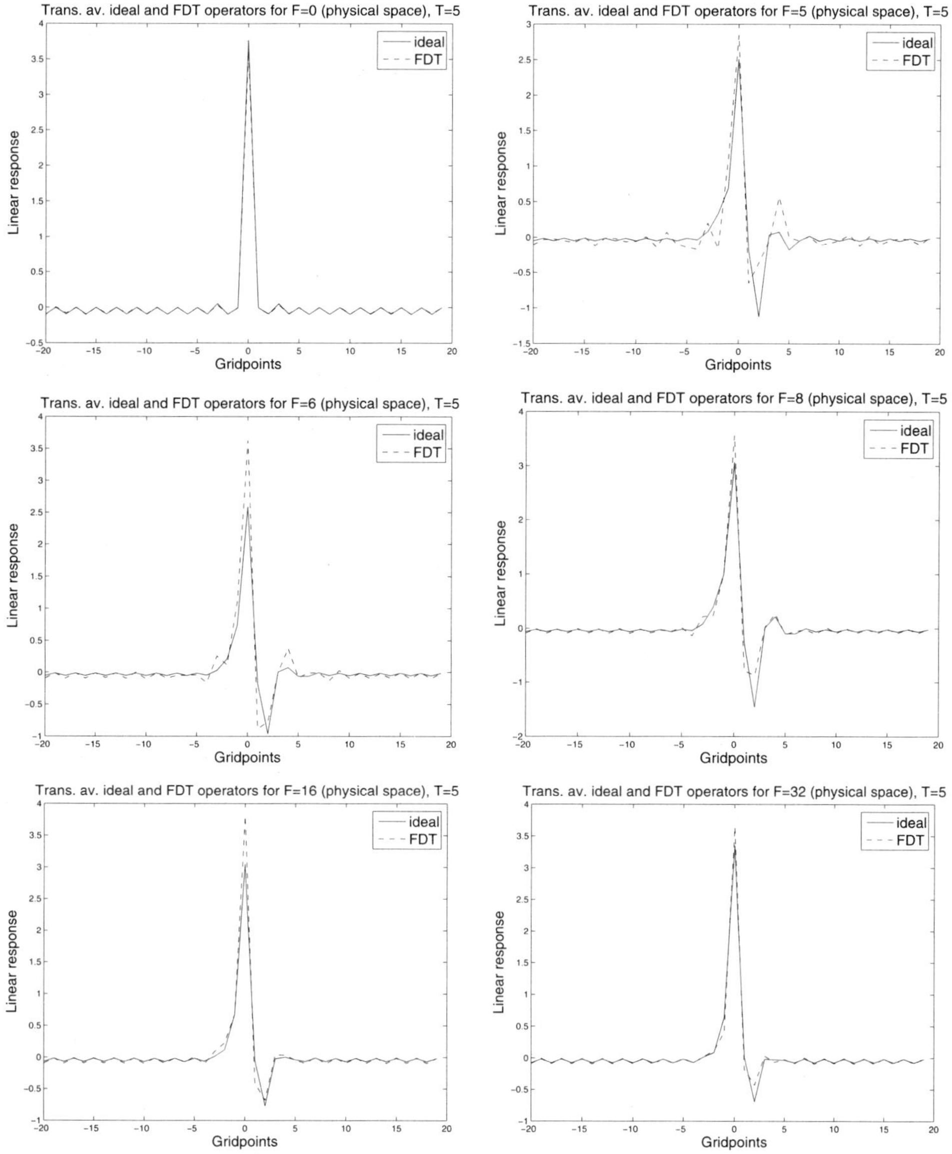

FIGURE 2.44. The translationally averaged linear response operator in physical space. Comparison between FDT and ideal response. L96 model, dynamical regimes $F = 0, 5, 6, 8, 16, 32$. Time $T = 5$.

for the values of forcing $F = 8, 16, 32$. Figure 2.46 shows the translationally averaged linear response operators in physical space for the values of forcing $F = 5$, 6, 8, 16, 32 as well as the undamped unforced regime with $F = 0$. Again, as at the time $T = 5$, one can see that the response for the undamped unforced regime $F = 0$ is essentially a δ-function at the perturbation gridpoint with small oscillations elsewhere, whereas responses for $F \neq 0$ show stranger asymmetric off-diagonal

TABLE 2.18. The errors and physical correlations for the regimes $F = 0, 5, 6, 8, 16, 32$, L96 model, at time $T = 20$. The results are displayed for real and imaginary parts of linear response operators in Fourier space, the translationally averaged linear response operators in physical space, and the spectra of singular values.

$F = 0$, time $T = 20$	L^2 error	$C(g,h)$
Fourier diag.	0.04666	0.9995
Phys. av.	0.03954	0.9998

$F = 5$, time $T = 20$	L^2 error	$C(g,h)$
Fourier diag.	0.8854	0.7408
Phys. av.	0.834	0.7392

$F = 6$, time $T = 20$	L^2 error	$C(g,h)$
Fourier diag.	0.7617	0.796
Phys. av.	0.7456	0.7866

$F = 8$, time $T = 20$	L^2 error	$C(g,h)$
Fourier diag.	0.6515	0.8092
Phys. av.	0.6417	0.8076

$F = 16$, time $T = 20$	L^2 error	$C(g,h)$
Fourier diag.	0.3782	0.9364
Phys. av.	0.4062	0.9279

$F = 32$, time $T = 20$	L^2 error	$C(g,h)$
Fourier diag.	0.246	0.9817
Phys. av.	0.289	0.9704

response. Note that this asymmetric response is stranger to the right of the perturbation and is consistent with the tendency of energy to go eastward in solutions, see Figure 2.34. The undamped unforced regime $F = 0$ shows perfect matching of ideal and FDT operators, however forced damped regimes have some discrepancies between the FDT and ideal response operators. It can be seen clearly that with increasing F the FDT approximation improves significantly. However, observe that even for the small values of forcing $F = 5, 6$ the oscillations of the ideal response are partially recovered by the FDT, again with significant overshoot like for the time $T = 5$. The effect of shifting the response towards smaller scales does not seem to have any impact for the response operators in physical space. In Table 2.18 we display L^2 relative error charts defined in (2.84) and physical correlations from (2.85). According to the Table 2.18 the FDT approximation improves a lot when F increases, with the best result being the undamped unforced case $F = 0$. Note that although the L^2 errors are large in some cases with small F, the correlations $C(g, h)$ are close to 1 in all cases (not as close as for the time $T = 5$), which means that the FDT is rather collinear with the ideal response, while consistently overshooting it. Furthermore, these correlations of the response operator improve as F increases.

2.8.3.4. *Linear response operators for the time* $T = \infty$. Here we show the properties of the linear response operators at the time $T = \infty$. While the quasi-Gaussian FDT approximation for $T = \infty$ is compute in our usual way, that is, by integrating the correlation functions from (2.81) for a very long time window such that they all decay to zero, the ideal infinite time response $\mathcal{R}_{\text{Ideal}}(\infty)$ is computed through long time averages of perturbed solutions rather than ensemble averages, following the procedure set forth in Section 2.5. Such a set-up is apparently not applicable for the unforced undamped regime $F = 0$ since the energy-preserving system can not be perturbed indefinitely by a constant forcing with no damping.

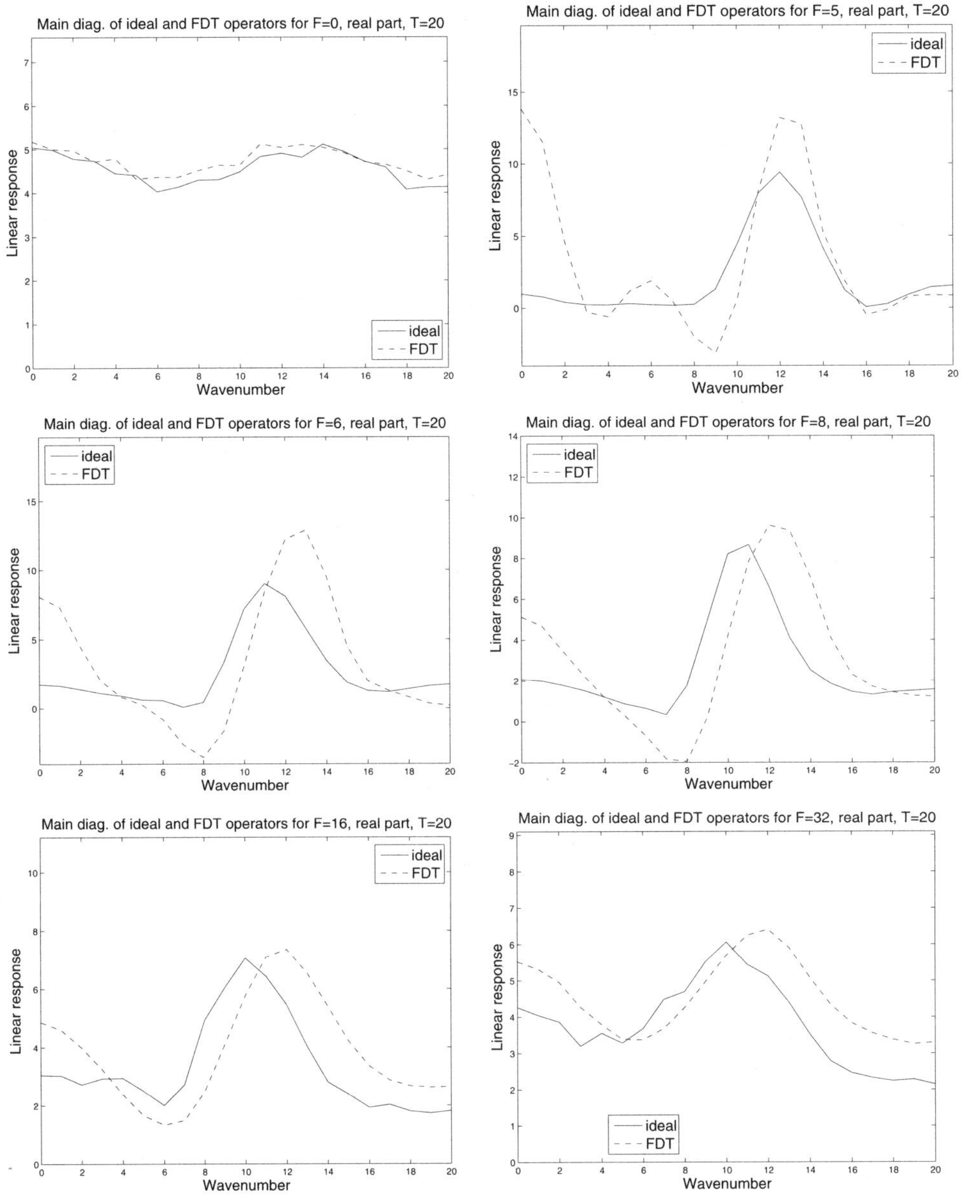

FIGURE 2.45. The real part of the main diagonal of the linear response operator in Fourier space. Comparison between FDT and ideal response. L96 model, dynamical regimes $F = 0, 5, 6, 8, 16, 32$. Time $T = 20$.

Thus, the results for the infinite time response are restricted to damped forced quasi-Gaussian FDT approximations with forcing set to F = 5, 6, 8, 16, 32. In Figure 2.47 we check the validity of the FDT approximations for the regimes F = 5, 6, 8, 16, 32 by showing the real parts of the main diagonals of both FDT and ideal linear response operators in the Fourier space. Again, as for the finite times $T = 5$

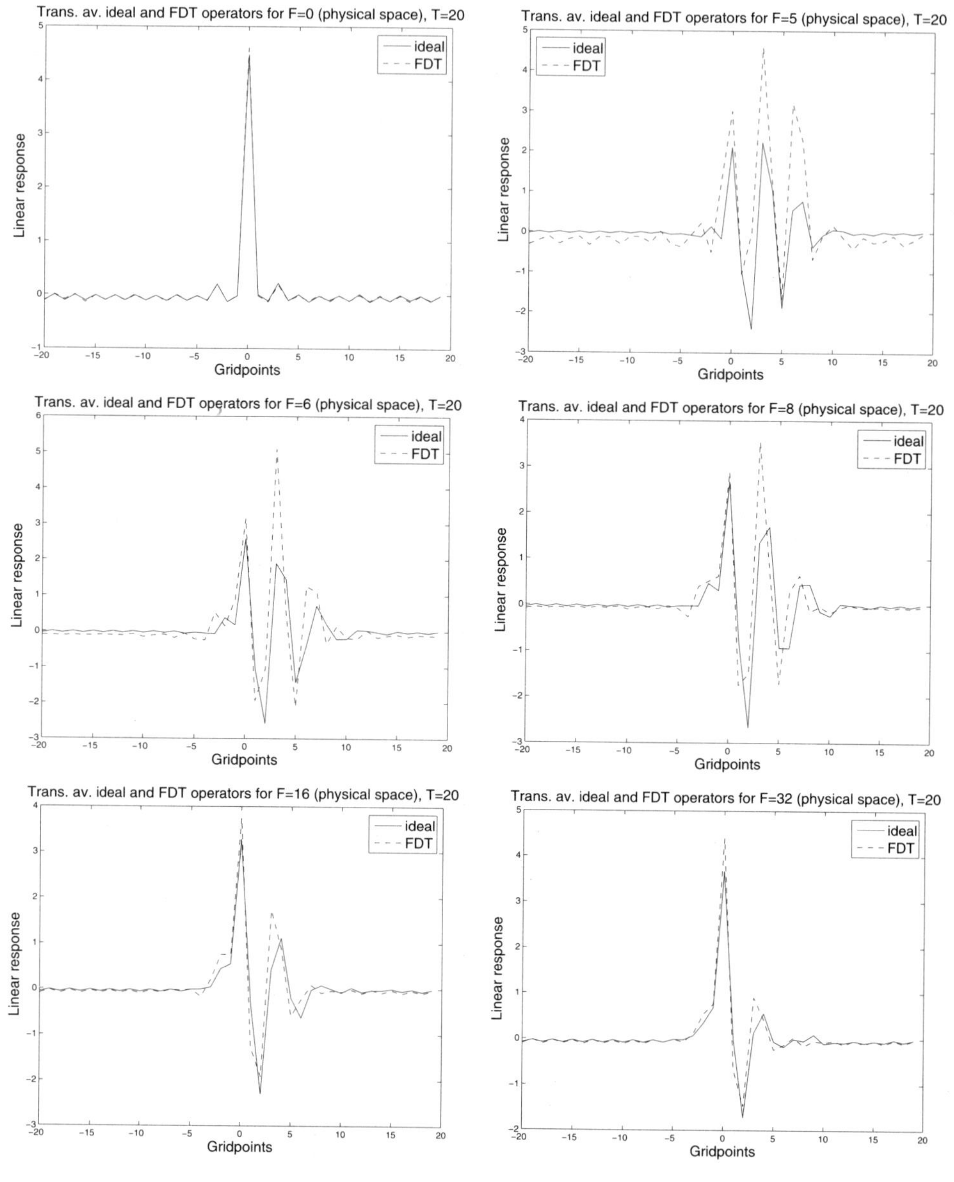

FIGURE 2.46. The translationally averaged linear response operator in physical space. Comparison between FDT and ideal response. L96 model, dynamical regimes $F = 0,\ 5,\ 6,\ 8,\ 16,\ 32$. Time $T = 20$.

and $T = 20$, the FDT consistently overshoots the ideal response. The qualitative behavior of the quasi-Gaussian FDT is, however, similar to that at time $T = 20$. Observe that the in general the FDT recovers the shape of the main diagonal of the ideal response operator, however for the regimes $F = 8, 16, 32$ it seems to be shifted towards smaller scales by one or two wavenumbers. No wavenumber shift occurs for regimes $F = 5,\ 6$. Figure 2.48 shows translationally averaged linear

TABLE 2.19. The errors and physical correlations for the regimes $F = 0, 5, 6, 8, 16, 32$, L96 model, at time $T = \infty$. The results are displayed for real and imaginary parts of linear response operators in Fourier space, the translationally averaged linear response operators in physical space, and the spectra of singular values.

$F = 5$, inf. resp.		
	L^2 error	$C(g,h)$
Fourier diag.	1.046	0.7066
Phys. av.	0.8408	0.7322

$F = 6$, inf. resp.		
	L^2 error	$C(g,h)$
Fourier diag.	0.7868	0.8228
Phys. av.	0.8157	0.7765

$F = 8$, inf. resp.		
	L^2 error	$C(g,h)$
Fourier diag.	0.7061	0.7761
Phys. av.	0.7135	0.7618

$F = 16$, inf. resp.		
	L^2 error	$C(g,h)$
Fourier diag.	0.4346	0.9159
Phys. av.	0.4493	0.9082

$F = 32$, inf. resp.		
	L^2 error	$C(g,h)$
Fourier diag.	0.3296	0.9574
Phys. av.	0.3411	0.9531

response operators in physical space for the values of forcing $F = 5, 6, 8, 16, 32$. It can be clearly seen that with increasing F the FDT approximation improves significantly. However, observe that even for small values of forcing $F = 5, 6$ the oscillations of the ideal response are partially recovered by the FDT, although with a huge overshoot. The effect of shifting the response towards smaller scales does not seem to have any impact for the response operators in physical space for the regimes $F = 8, 16, 32$. In Table 2.19 we display L^2 relative error charts defined in (2.84) and physical correlations from (2.85). According to the Table 2.19 the FDT approximation improves a lot when F increases. The L^2 error decreases from $0.8 - 1$ for $F = 5$ to 0.3 for $F = 32$. The collinearity $C(g,h)$ increases from 0.7 for the value of forcing $F = 5$ to 0.95 for $F = 32$. Note once again the highly asymmetric wave packets in the response operator at various times.

2.8.4. Summary. Here we discuss the results of the quasi-Gaussian FDT for the L96 model. Overall, it turns out that despite the presence of damping and forcing with no explicit energy conservation, the quasi-Gaussian FDT approximation can be very efficient for prediction of climate response to small external forcing, although it is not as accurate as the FDT approximation for the energy-conserving chaotic models with no forcing and damping like TBH, K-Z and IL96, described in Section 2.6. Below we summarize the results and trends for the L96 model in more detail.

The energy based scaling argument in (2.117) seems to be appropriate for both the climatology and the quasi-Gaussian FDT for the L96 model. The main trend observed was that the main statistical properties of the damped forced L96 model in the rescaled coordinate-time frame given in (2.115) indeed systematically approach the undamped unforced IL96 system. This trend is observed to be valid for the variance spectrum, time autocorrelation functions and the correlation time

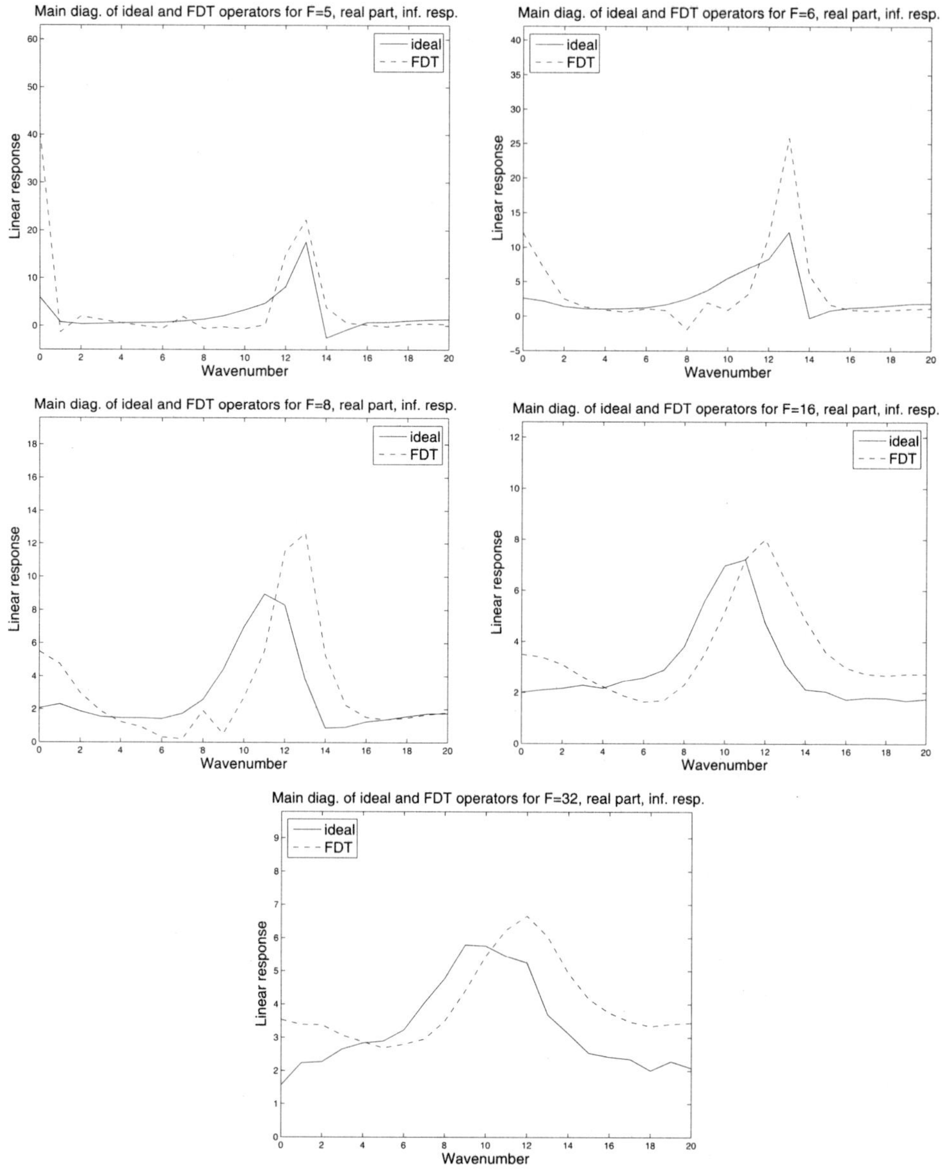

FIGURE 2.47. The real part of the main diagonal of the linear response operator in Fourier space. Comparison between FDT and ideal response. L96 model, dynamical regimes $F = 0$, 5, 6, 8, 16, 32. Time $T = \infty$.

spectrum with few exceptions. One exception is a part of the spectrum $13 \leq k \leq 20$ which is linearly stable for all values of F. This part of the spectrum seems to lack the convergence towards the undamped unforced dynamics as F increases. For the time autocorrelation functions, there is a common time interval ($T = 5 - 10$, depending on a wavenumber) where the correlation functions are

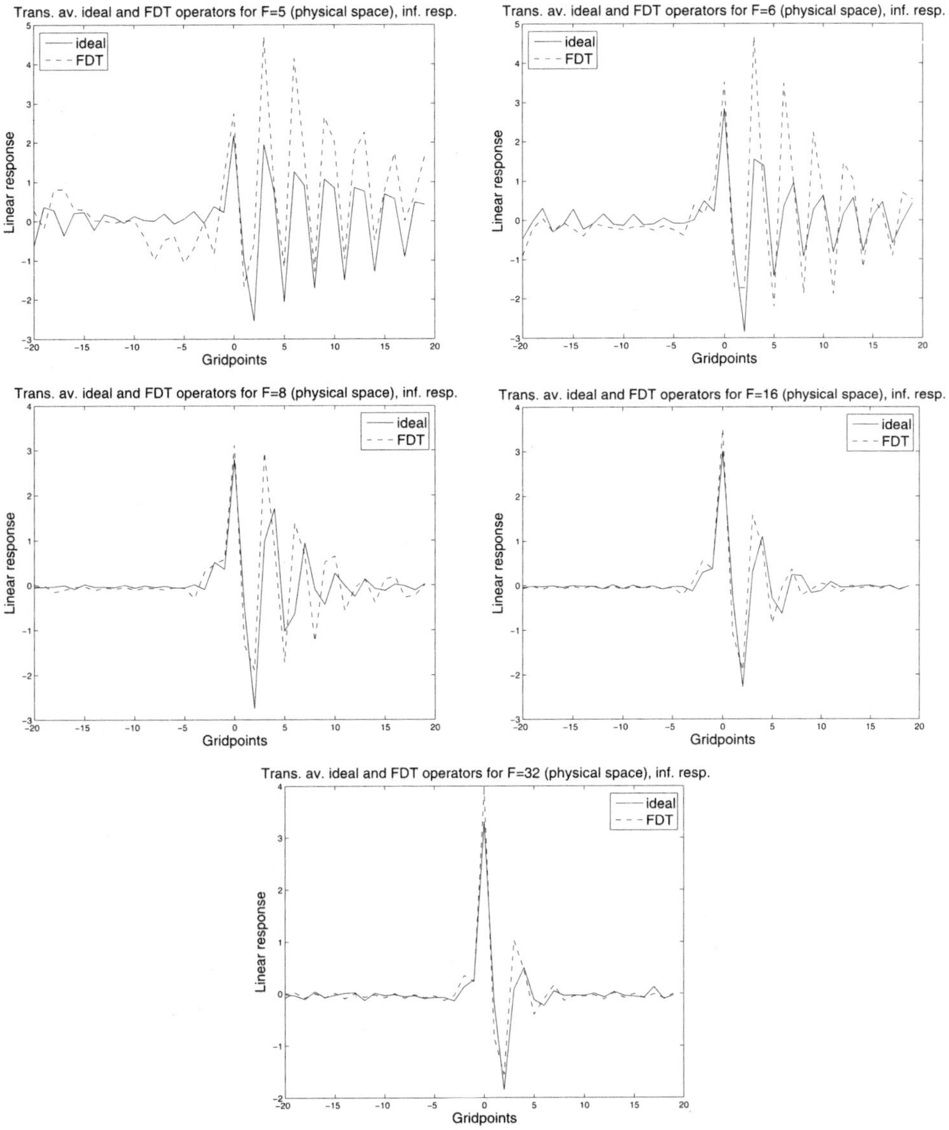

FIGURE 2.48. The translationally averaged linear response operator in physical space. Comparison between FDT and ideal response. L96 model, dynamical regimes $F = 0,\ 5,\ 6,\ 8,\ 16,\ 32$. Time $T = \infty$.

self-similar. After that time, the correlation function for the regimes with low F become slowly decaying and sometimes highly oscillatory, whereas the fast mixing regimes for high $F = 16, 32$ approach the unforced undamped IL96 regime. The singular values for the FDT Gaussian linear response operators obey essentially the same trend as the correlation functions because the FDT quasi-Gaussian operators are the time integrals of the time correlation functions. The validity of the quasi-Gaussian FDT for singular values extends up to $T = 20$ for the regimes $F = 8$,

16, 32. For the snapshots of linear response operators at times $T = 5$, 20 and for the infinite time response operators, signs of convergence of damped forced regimes to the undamped unforced regime in rescaled coordinates are observed as F increases. The inner product error norms and physical correlations (collinearity test) show strong convergence of the damped forced regimes towards the undamped unforced regime as forcing increases. In particular, the collinearity for regimes with moderate and high F can be almost perfect even though the error norm shows some discrepancy. Additionally, even for the cases where the quasi-Gaussian FDT can no longer be considered valid due to high L^2 error, we can observe that ideal response profile in physical space is captured by FDT qualitatively correctly, which means that the FDT approximation exhibits similar trends (the peaks, valleys and oscillations are all in the right places), however the magnitude of those captured features is nowhere near the ideal measured response. This kind of "qualitative validity" can sometimes be useful in practice for long-term climate prediction, even though the correct response is not captured well. For all values of $F \neq 0$ the quasi-Gaussian approximation for FDT has the largest errors for the stable wavenumbers, $0 \leq k \leq 3$ and $13 \leq k \leq 20$, where this approximation overshoots the ideal response.

2.9. Nonlinear functionals and FDT

While the linear functional $\vec{A}(\vec{X}) = \vec{X}$ is the simplest physically relevant quantity to study for its linear response, the fluctuation-dissipation theorem as described in Section 2.2 provides the explicit linear response formula for an arbitrary functional $\vec{A}(\vec{X})$. It has been already demonstrated in Section 2.6 that the FDT for the linear functional $\vec{A}(\vec{X}) = \vec{X}$ works perfectly for the chaotic energy-conservative systems with Gibbs equilibrium, like TBH, K-Z and IL96 models. On the other hand, while $\vec{X}$ generally represents physically important quantity like mass or momentum, there are quantities which are equally important and nonlinear at the same time, like the energy. In particular, the linear response of energy density, energy per gridpoint, or the variance,

$$A_i(\vec{X}) = X_i^2, \tag{2.127}$$

to small external perturbations is of particular interest in predictions of long-term climate change. In this section we test the response of the quadratic functional in (2.127) together with the cubic nonlinear functional,

$$A_i(\vec{X}) = X_i^3, \tag{2.128}$$

for the K-Z model, given by (2.76). The formula (2.34) for constant external forcing δF provides the linear response operator of the form

$$\mathcal{R}_{ij,\mathrm{FDT}}(t) = \theta \int_0^t \langle X_i^2(\tau + t_0) X_j(t_0) \rangle \, d\tau \tag{2.129}$$

for the quadratic functional in (2.127), and

$$\mathcal{R}_{ij,\mathrm{FDT}}(t) = \theta \int_0^t \langle X_i^3(\tau + t_0) X_j(t_0) \rangle \, d\tau \tag{2.130}$$

for (2.128), respectively, where X_i are now physical space gridpoints, unlike the experiments in Sections 2.6 and 2.8 which were done in Fourier space. The ideal response is computed in the same way as described in Section 2.5, but individual physical gridpoints are now perturbed instead of Fourier modes. We pick the K-Z model in (2.76) to test the validity of FDT for the nonlinear functionals in (2.127) and (2.128) due to the fact that it exhibits rich bifurcating structure in the response at various spatial scales as t varies and simultaneously has a simple finite-difference type of representation in physical space, thus combining easy implementation and a rich physical pattern of response to small external perturbations.

2.9.1. The validity of FDT for quadratic response functional. In this section we show the results of FDT validation for the linear response of the quadratic energy density functional in (2.127). Following the strategy set forth in Section 2.6, we first validate the time evolution of few selected singular values of the linear response operator, and follow with more detailed study of the response at instantaneous time slices $T = 5$ and $T = 20$.

2.9.1.1. *Time evolution of singular values.* In Figure 2.49 we show the time evolution of the four largest singular values for the K-Z model and the nonlinear response functional $A_i(\vec{X}) = X_i^2$. Observe that the FDT approximations for singular values 1, 2 and 4 closely follow the ideal measurement, whereas the FDT for the singular value 3 stops being valid at about time $T = 15$. Starting with time $T = 15$, the FDT approximation for the third singular value of the response matrix consistently overshoots the ideal measured response.

2.9.1.2. *Linear response operators for the time $T = 5$.* Here we show the properties of the linear response operators at the time $T = 5$. Figure 2.52 shows translationally averaged linear response operators in physical space. Observe that for the quadratic energy density functional in (2.127) the response is characterized by an odd function in physical space, contrary to the response of a linear functional studied in Section 2.6. In particular, there is no response at the gridpoint where the small external forcing is applied (whereas for a linear functional the response is maximal at the same point). We can see that for the time $T = 5$ the FDT captures this remarkable response shape quite well. In Figure 2.53 we present the spectra of singular values for the linear response operator at the time $T = 5$. The spectrum of singular values shows a good match of the FDT approximation with the ideal measurement. In Table 2.20 we display L^2 relative error chart defined in (2.84) and physical correlation from (2.85). The L^2 error is larger than the errors for a linear functional from Section 2.6 for the same K-Z model; however it does not exceed 20%, which can be considered a good match for long term climate response predictions, at least at their current stage. The physical correlation of the ideal response and its FDT approximation is also not as good as for a linear functional in Section 2.6 for the K-Z model, but is still very good $\big(C(g,h) = 0.98)\big)$.

2.9.1.3. *Linear response operators for the time $T = 20$.* Here we show the properties of the linear response operators at the time $T = 20$. Figure 2.50 shows

TABLE 2.20. The error and physical correlation for the K-Z model and nonlinear functional $A_i(\vec{X}) = X_i^2$ at time $T = 5$. The results are displayed for the translationally averaged linear response operators in physical space.

$A_i = X_i^2$, Time $T = 5$		
	L^2 error	$C(g,h)$
Phys. av.	0.1984	0.9804

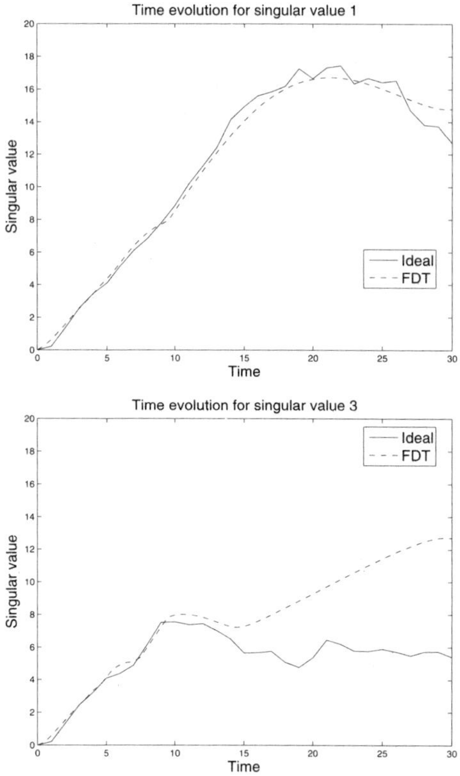

FIGURE 2.49. The time evolution for singular values 1–4, K-Z model, nonlinear functional $A_i(\vec{X}) = X_i^2$.

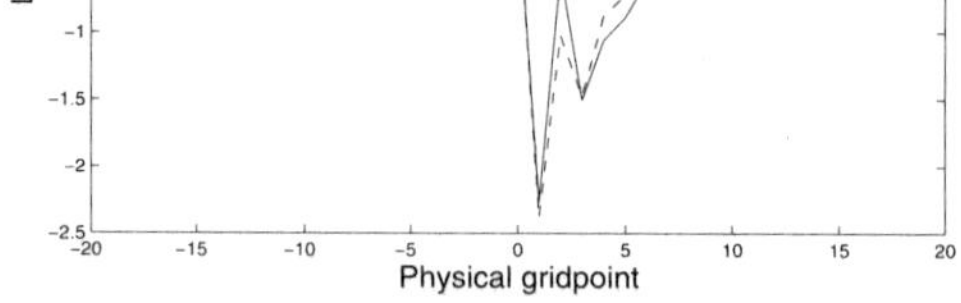

FIGURE 2.50. The translationally averaged linear response operator in physical space for K-Z model and nonlinear functional $A_i(\vec{X}) = X_i^2$. Time $T = 5$.

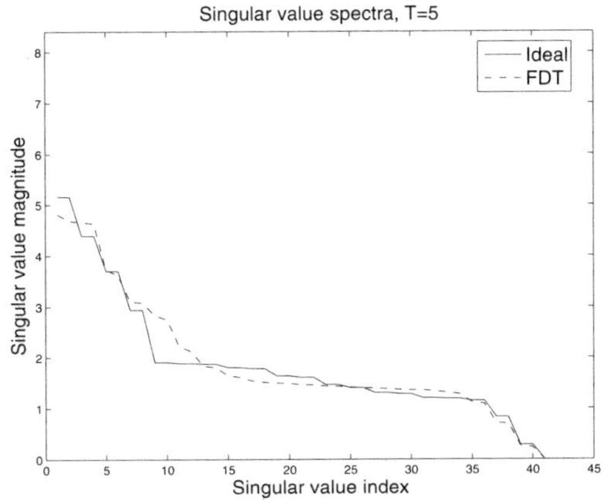

FIGURE 2.51. The spectrum of singular values for linear response operators for K-Z model and nonlinear functional $A_i(\vec{X}) = X_i^2$. Time $T = 5$.

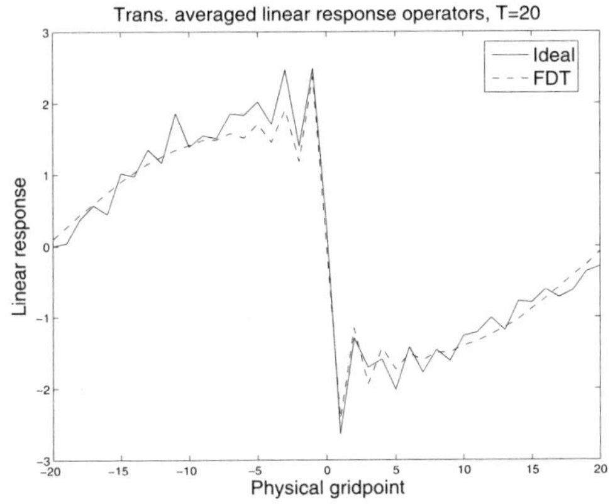

FIGURE 2.52. The translationally averaged linear response operator in physical space for K-Z model and nonlinear functional $A_i(\vec{X}) = X_i^2$. Time $T = 20$.

translationally averaged linear response operators in physical space. Observe that for the time $T = 20$ the FDT approximation captures the response shape with roughly same precision as for the time $T = 5$. In Figure 2.51 we present the spectra of singular values for the linear response operator at the time $T = 20$. The spectrum of singular values shows almost a perfect match of the FDT approximation with the ideal measurement. In Table 2.21 we display L^2 relative error chart defined in (2.84) and the physical correlation from (2.85). The L^2 error is, in fact, smaller than that for the time $T = 5$ (16% vs 20%), however it is larger than errors for a linear functional from Section 2.6 for the same K-Z model. The physical correlation of the ideal response and its FDT approximation is also better than that for the time $T = 5$ (0.99 vs 0.98), but not as good as for a linear functional in Section 2.6 for the K-Z model.

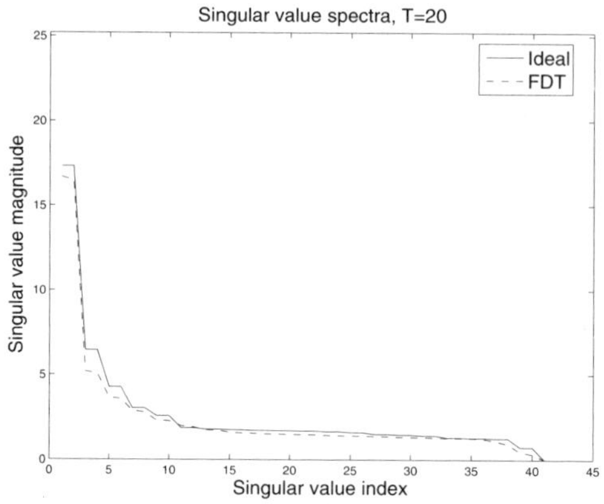

FIGURE 2.53. The spectrum of singular values for linear response operators for K-Z model and nonlinear functional $A_i(\vec{X}) = X_i^2$. Time $T = 20$.

TABLE 2.21. The error and physical correlation for the K-Z model and nonlinear functional $A_i(\vec{X}) = X_i^2$ at time $T = 20$. The results are displayed for the translationally averaged linear response operators in physical space.

$A_i = X_i^2$, Time $T = 20$		
	L^2 error	$C(g, h)$
Phys. av.	0.1582	0.99

2.9.2. The validity of FDT for cubic response functional. In this section we show the results of FDT validation for the linear response of the cubic functional in (2.128). Following the strategy set forth in Section 2.6, we first validate the time evolution of few selected singular values of the linear response operator, and follow with more detailed study of the response at instantaneous time slices $T = 5$ and $T = 20$.

2.9.2.1. *Time evolution of singular values.* In Figure 2.54 we show the time evolution of four largest singular values for the K-Z model and the nonlinear cubic response functional $A_i(\vec{X}) = X_i^3$. Observe that here the FDT can not be possibly considered a good approximation to the linear response, since the ideal singular values overshoot the FDT predictions by roughly a factor of 2. Note, however, that the qualitative behavior of the FDT time evolution captures major features of the ideal response, that is, the FDT singular values, if properly scaled with height, would be quite similar to the ideal response.

2.9.2.2. *Linear response operators for the time $T = 5$.* Here we show the properties of the linear response operators at the time $T = 5$. Figure 2.55 shows translationally averaged linear response operators in physical space. Observe that for the nonlinear cubic functional in (2.128) the response is characterized by a strong peak at the gridpoint of small external perturbation with oscillations around it, similar to the response of a linear functional studied in Section 2.6, and contrary to the response profile of the quadratic energy density functional studied above. We can

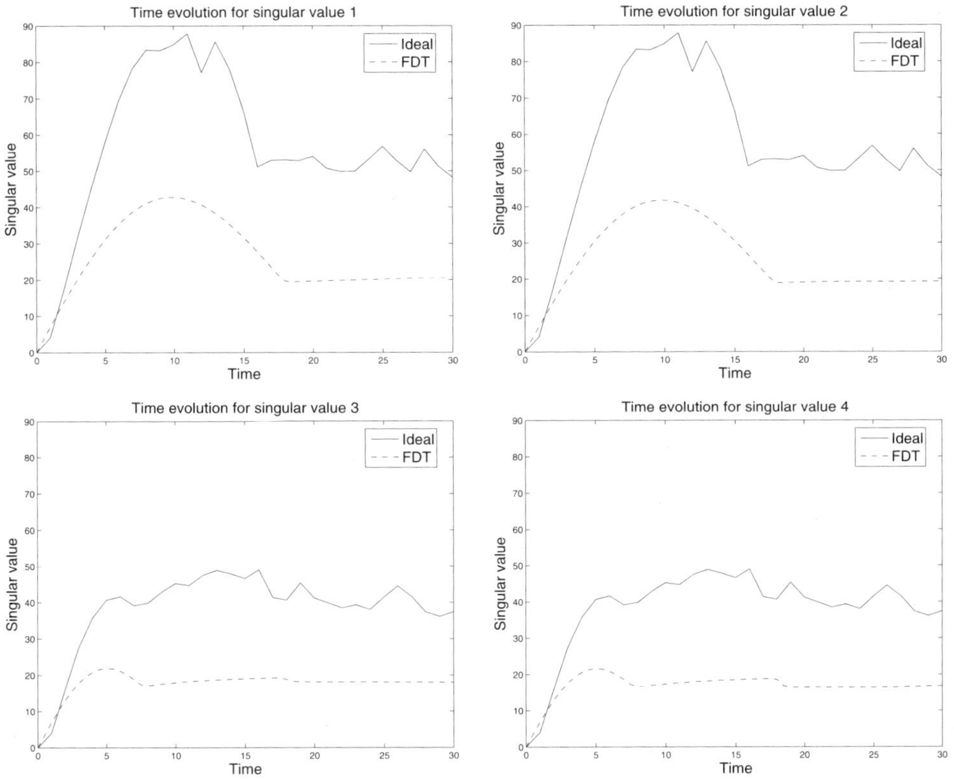

FIGURE 2.54. The time evolution for singular values 1–4, K-Z model, nonlinear functional $A_i(\vec{X}) = X_i^3$.

TABLE 2.22. The error and physical correlation for the K-Z model and nonlinear functional $A_i(\vec{X}) = X_i^3$ at time $T = 5$. The results are displayed for the translationally averaged linear response operators in physical space.

$A_i = X_i^3$, Time $T = 5$		
	L^2 error	$C(g,h)$
Phys. av.	0.7644	0.9938

see that for the time $T = 5$ the ideal response systematically undershoots the FDT approximation by roughly a factor of 2, although the latter captures the shape and qualitative features of the ideal response. In Figure 2.56 we present the spectra of singular values for the linear response operator at the time $T = 5$. Again, the FDT spectrum would be a good match for the ideal one if scaled by roughly a factor of 2. In Table 2.22 we display L^2 relative error chart defined in (2.84) and physical correlation from (2.85). The L^2 error here is 76% which is apparently huge. Remarkably, the physical correlation of the ideal response and its FDT approximation is nearly perfect $\big(C(g,h) = 0.99)\big)$, which means that the ideal response and FDT approximation are collinear, but have different magnitudes.

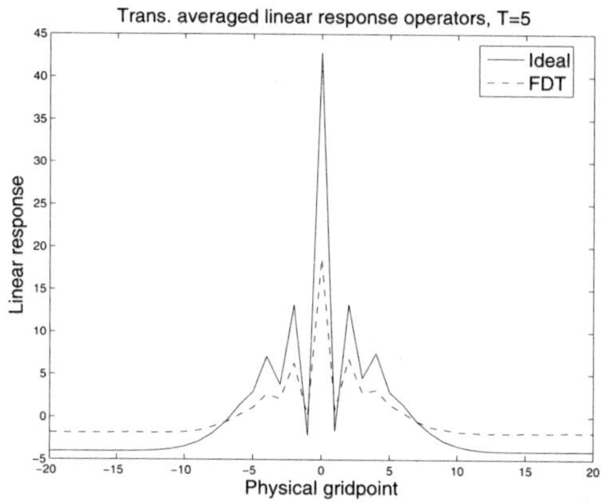

FIGURE 2.55. The translationally averaged linear response operator in physical space for K-Z model and nonlinear functional $A_i(\vec{X}) = X_i^3$. Time $T = 5$.

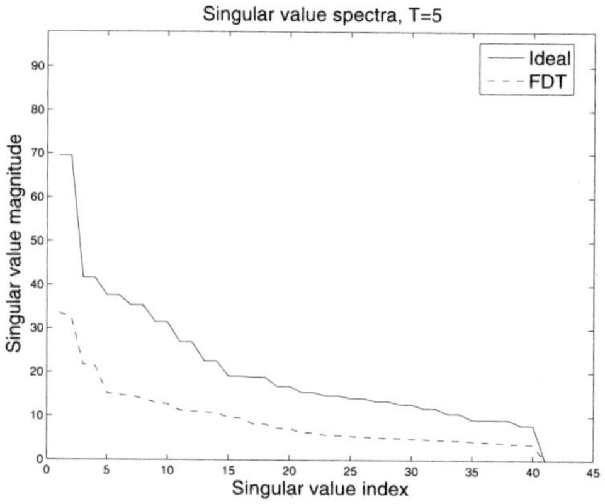

FIGURE 2.56. The spectrum of singular values for linear response operators for K-Z model and nonlinear functional $A_i(\vec{X}) = X_i^3$. Time $T = 5$.

2.9.2.3. *Linear response operators for the time* $T = 20$. Here we show the properties of the linear response operators at the time $T = 20$. Figure 2.57 shows translationally averaged linear response operators in physical space. Observe that for the time $T = 20$ the FDT approximation is as bad as for the time $T = 5$, nonetheless capturing the shape and bulk features of the ideal response. In Figure 2.55 we present the spectra of singular values for the linear response operator at the time $T = 20$. Again, like for the time $T = 5$, the ideal spectrum of singular values overshoots its FDT approximation by roughly a factor of 2. In Table 2.21 we display L^2 relative error chart defined in (2.84) and physical correlation from (2.85). The situation here is also very similar to the time $T = 5$: while the L^2 error is about 90%, the physical correlation $C(g, h) = 0.98$ which is nearly perfect. This means that although the FDT approximation misses the ideal response by far in magnitude, both responses are highly collinear in shape.

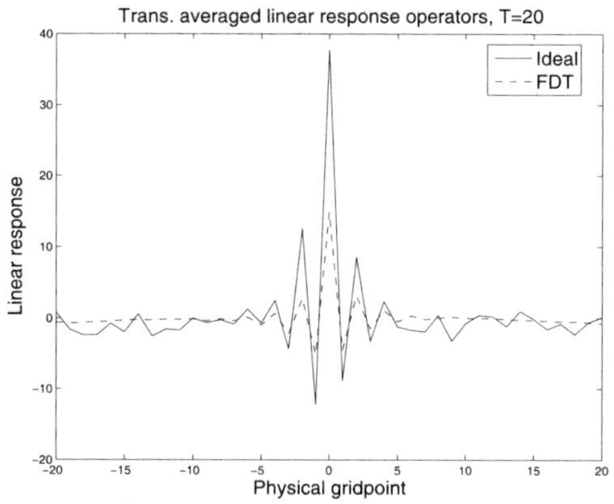

FIGURE 2.57. The translationally averaged linear response operator in physical space for K-Z model and nonlinear functional $A_i(\vec{X}) = X_i^3$. Time $T = 20$.

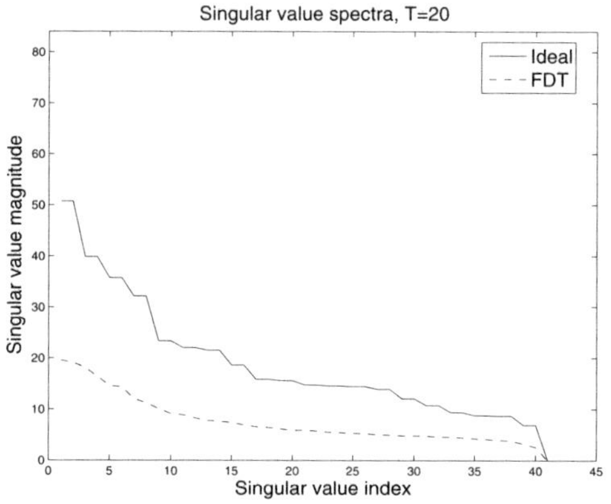

FIGURE 2.58. The spectrum of singular values for linear response operators for K-Z model and nonlinear functional $A_i(\vec{X}) = X_i^3$. Time $T = 20$.

TABLE 2.23. The error and physical correlation for the K-Z model and nonlinear functional $A_i(\vec{X}) = X_i^3$ at time $T = 20$. The results are displayed for the translationally averaged linear response operators in physical space.

$A_i = X_i^3$, Time $T = 20$		
	L^2 error	$C(g,h)$
Phys. av.	0.9027	0.9765

2.9.3. Summary. Here we discuss the results of the FDT approximation for the two nonlinear functionals in physical space, quadratic (2.127) and cubic (2.128), and the K-Z model given in (2.76). The remarkable property of the response for the quadratic energy density functional is that its physical space profile is apparently an odd function with respect to the external perturbation. In particular, the response is zero at the gridpoint of external perturbation, contrary to the response of linear functional studied in Section 2.6. Despite this nontrivial response profile, the FDT approximation reproduces it quite well: the L^2 errors do not exceed 20% for the time range considered, and the collinearity of the FDT with the response is about 0.98–0.99, which is a very high value. On the other hand, the response of the cubic functional to a small external forcing is characterized by the same properties as the one for the trivial linear functional studied in Section 2.6, including a peak at the gridpoint of small external perturbation surrounded by oscillations, and it turns out that the FDT is not valid for this cubic functional in a strict sense. The remarkable observation is, however, that although the errors in magnitude for the physical space profile of the response are 75–90% (the ideal response overshoots its FDT approximation by roughly a factor of 2), the collinearity $C(g, h)$ of the FDT and ideal responses is as high as for the quadratic response functional, $C(g, h) =$ 0.98–0.99, which means that, with appropriate (presumably heuristic) amplitude rescaling, the FDT approximation can be tailored to match the ideal measurement quite well.

2.10. Future research directions for FDT in complex nonlinear systems

Given the computational studies regarding the validity of the classical FDT reported above in Sections 2.6, 2.8 and 2.9, the outstanding practical problem remaining is to improve the FDT approximation beyond the quasi-Gaussian approximation of Section 2.7 for deterministic forced dissipative dynamical systems with many degrees of freedom, the structure in (2.94), and the anisotropy associated with realistic geophysical systems. The major difficulty in this situation is that the measure $p_{\mathrm{eq}}(\vec{X})$ in this case is typically an SRB measure which is supported on a large-dimensional (often fractal) set and is usually not absolutely continuous with respect to the Lebesgue measure [**5, 24**]. In the context of Axiom A attractors, Ruelle [**21, 22**] has generalized the classical calculations for FDT presented above in Section 2.2 to this setting but it is far from obvious how to use these formulas in a computational framework. For example, for special functionals $A(\vec{X})$ which reside on the expanding directions of the attractor, the Axiom A FDT formulas are a modified version of those from Section 2.2 while for functionals $A(\vec{X})$ which also project on the contracting dissipative directions, the Axiom A FDT formulas are more complex.

In the results presented in Section 2.8 for the L96 model, the quasi-Gaussian FDT approximation systematically overestimates the ideal response and this is true especially for the linearly stable dissipative directions in phase space as established there; a similar (sometimes large!) consistent overestimate of the response occurs in other tests of the quasi-Gaussian approximation for FDT in this dissipative context [**2, 8, 9**]. We have established in Section 2.8 that at least for the L96 model, this overestimate of the response decreases as F increases for values of $F = 8$, 16, 32; the key idea there was an a posteriori rescaling of the dynamical system as F increased so that the vector field defining the dynamics becomes formally weakly convergent and, at least according to linear theory, the contracting directions become weakly dissipative. In these circumstances we established in Section 2.8 that the quasi-Gaussian FDT approximation to the ideal response improves, confirming our heuristic intuition. The first two authors are currently investigating possible improvements of the FDT using the more sophisticated machinery of the Axiom A FDT. Also it should be mentioned here that Gallavotti and Cohen [**7**] have developed an interesting generalization of FDT for dissipative systems with time reversal symmetry (strongly violated in geophysical models) with interesting applications to thermostated systems of identical particles.

An alternative route to improving the quasi-Gaussian FDT approximation is suggested by Theorem 2.3 in Section 2.7 which establishes the surprisingly high accuracy of this approximation at short times for special functionals. Clearly, using higher moment estimators as discussed in Chapter 1 will improve this approximation but practical computationally feasible strategies for doing this in the large phase space are needed such as those in [**1, 14**]. It is especially interesting to develop an improved practical strategy beyond the quasi-Gaussian approximation for the response to steady forcing. The first two authors also are currently investigating this avenue of research.

We have intentionally presented the FDT Theorem 2.1 in Section 2.2 in the context of stochastically forced nonlinear dynamical systems. We have done this for several reasons. First, for theoretical reasons, the introduction of appropriate stochastic noise in a dynamical system typically yields a unique smooth rapidly decreasing invariant density $p_{\mathrm{eq}}(\vec{X})$ so the formal FDT calculations can be rigorously justified; a standard route in dynamical system theory is to approximate the SRB measure of a deterministic dynamical system as the limit of $p_{\mathrm{eq}}(\vec{X})$ with small noise as $\varepsilon \to 0$ [**24**]. Secondly, for many practical systems such as occur in the geosciences, the stochastic noise represents unresolved physical degrees of freedom. We provide an introduction to this important topic in Chapter 3.

Finally, establishing the mathematically rigorous validity of the FDT theorem for suitable functionals for long time is a challenge for mathematical analysis. The following example shows that this is a highly nontrivial issue. Referring back to the conservative models in (2.29) from Section 2.2, we perturb this system by a small external perturbation involving constant dissipation so that from (2.37)

$$\frac{L^i_{\mathrm{ext}} p_{\mathrm{eq}}}{p_{\mathrm{eq}}} = \bar{d}(-1 + \theta X_i^2). \tag{2.131}$$

Assume time-independent forcing and compute the linear response of $A(\vec{X}) = X_i^2$; according to the FDT,

$$\mathcal{R}_{\mathrm{FDT}}[X_i^2](T) = \delta\bar{d}\Big(- \langle X_i^2 \rangle T + \theta \int_0^T \langle X_i^2(t') X_i^2(0) \rangle \, dt' \Big). \tag{2.132}$$

Clearly the linear response, $\mathcal{R}_{\mathrm{FDT}}(X_i^2)$, has an unavoidable growing secular term $-\delta\bar{d}\langle X_i^2 \rangle T$, suggesting clearly that FDT can not be valid for all time for $\langle X_i^2 \rangle$. On the other hand, $\langle X_i \rangle = 0$ and for the linear functional, $A(\vec{X}) = X_i$,

$$\mathcal{R}_{\mathrm{FDT}}[X_i](T) = \delta\bar{d}\theta \int_0^T \langle X_i(t') X_i^2(0) \rangle \, dt', \tag{2.133}$$

suggesting the potential validity of FDT for all times for this special functional for a mixing system.

References

1. R. Abramov and A. J. Majda, *Quantifying uncertainty for non-Gaussian ensembles in complex systems*, SIAM J. Sci. Comput. **26** (2004), no. 2, 411–447.
2. T. Bell, *Climate sensitivity from fluctuation dissipation: Some simple model tests*, J. Atmospheric Sci. **37** (1980), no. 8, 1700–1708.
3. G. F. Carnevale, M. Falcioni, S. Isola, R. Purini, and A. Vulpiani, *Fluctuation-response in systems with chaotic behavior*, Phys. Fluids A **3** (1991), no. 9, 2247–2254.
4. B. Cohen and G. Craig, *The response time of a convective cloud ensemble to a change in forcing*, The Royal Meteorological Society **130** (2004), no. 598, 933–944.
5. J. P Eckmann and D. Ruelle, *Ergodic theory of chaos and strange attractors*, Rev. Modern Phys. **57** (1985), no. 3, 617–656.
6. D. J. Evans and G. P. Morriss, *Statistical mechanics of nonequilibrium liquids*, Academic Press, New York, 1990.
7. G. Gallavotti and E. G. D. Cohen, *Dynamical ensembles in nonequilibrium statistical mechanics*, Phys. Rev. Lett. **74** (1995), 2694–2697.
8. A. S. Gritsoun, *Fluctuation-dissipation theorem on attractors of atmospheric models*, Russian J. Numer. Anal. Math. Modelling **16** (2001), no. 2, 115–133.
9. A. S. Gritsoun and V. P. Dymnikov, *Barotropic atmosphere response to small external actions. theory and numerical experiments*, zv. Russ. Acad. Sci. Atmospher. Ocean. Phys. **35** (1999), no. 5, 511–525.
10. S. Gritsoun, G. Branstator, and V. P. Dymnikov, *Construction of the linear response operator of an atmospheric general circulation model to small external forcing*, Russian J. Numer. Anal. Math. Modelling **17** (2002), 399–416.
11. K. Haven, A. J. Majda, and R. Abramov, *Quantifying predictability through information theory: Small sample estimation in a non-Gaussian framework*, J. Comput. Phys. (2004), in press.
12. R. Kubo, M. Toda, and N. Hashitsume, *Statistical physics.* II. *nonequilibrium statistical mechanics*, Springer-Verlag, New York, 1985.
13. C. E. Leith, *Climate response and fluctuation-dissipation*, J. Atmospheric Sci. **32** (1975), 2022–2025.
14. A. J. Majda, R. Kleeman, and D. Cai, *A framework for predictability through relative entropy*, Methods Appl. Anal. **9** (2002), 425–444.
15. A. J. Majda and I. Timofeyev, *Remarkable statistical behavior for truncated Burgers-Hopf dynamics*, Proc. Nat. Acad. Sci. **97** (2000), no. 23, 12413–12417.
16. ______, *Statistical mechanics for truncations of the Burgers–Hopf equation: A model for intrinsic stochastic behavior with scaling*, Milan J. Math. **70** (2002), 39–96.
17. A. J. Majda, I. Timofeyev, and E. Vanden-Eijnden, *A mathematical framework for stochastic climate models*, Comm. Pure Appl. Math. **54** (2001), no. 8, 891–974.
18. A. J. Majda and X. Wang, *Nonlinear dynamics and statistical theories for basic geophysical flows*, Cambridge University Press, New York, 2005.
19. J. Pedlosky, *Waves in the ocean and atmosphere*, Springer-Verlag, New York, 2003.
20. F. Risken, *The Fokker–Planck equation*, 2nd ed., Springer-Verlag, New York, 1988.
21. D. Ruelle, *Differentiation of SRB states*, Comm. Math. Phys. **187** (1997), 227–241.
22. ______, *General linear response formula in statistical mechanics, and the fluctuation-dissipation theorem far from equilibrium*, Phys. Lett. A **245** (1998), 220–224.
23. D. Williams, *Weighing the odds: A course in probability and statistics*, Cambridge University Press, New York, 2001.
24. L.-S. Young, *What are SRB measures, and which dynamical systems have them?*, J. Stat. Phys. **108** (2002), no. 5-6, 733–754.

CHAPTER 3

Mathematical Strategies for Stochastic Mode Reduction in Climate

3.1. Introduction to stochastic climate modeling

The climate system has a wide range of time scales for important physical processes. Thus on hourly scales there are boundary layer and small scale convective processes; on daily scales there are organized synoptic scale weather events; on monthly scales there is low frequency variability of the extra-tropics and intraseasonal variability of the tropics [**21**]; on annual scales there is the El Niño/Southern Oscillation and midlatitude planetary teleconnections [**44**]; finally, on decadal scales there are important oscillations of the coupled mid-latitude atmosphere/ocean system [**22**]. For example, in this last case, a reduced description of the atmosphere is needed in studying patterns of coupled decadal variability such as the North Atlantic Oscillation. There is an obvious need in such a complex system for systematic and rigorous strategies for reducing the number of degrees of freedom while retaining fidelity of important features of the unresolved variables. Such reductive strategies help to develop deeper physical understanding of the various components of the climate system and their interaction. They also provide useful reduced mathematical models that are much more computationally efficient. Hasselmann [**23**] proposed the idea of simplified stochastic modeling for the unresolved degrees of freedom in climate dynamics.

In recent years, stochastic modeling of the extra-tropical atmosphere has been a very active research topic [**4, 11, 43, 42, 5, 47, 1**], including stochastic modeling for the the angular momentum budget [**46, 15, 16**]. For example, as regards low frequency variability of the extra-tropical atmosphere, Branstator [**4**] and Whitaker and Sardeshmukh [**47**] have developed reduced linear stochastic models through regression fitting of linear Langevin dynamics, which assumes a structure with white noise forcing and linear damping. For the angular momentum budget, regression fitting of interesting phenomenological linear Langevin models [**46**] and higher order Markov models [**15, 16**] has been utilized. This stochastic modeling approach is the subject of a recent survey article [**10**].

A different approach to understanding low frequency variability of the atmosphere through the nonlinear dynamics and bifurcation theory of a reduced number of suitable climate variables has been developed beginning with the seminal work of Charney and DeVore [**6**]. This approach involves identifying structures such as multiple equilibria or stable periodic orbits in the phase space of reduced nonlinear climate variables [**32, 25, 21, 7**] and then utilizing the observational record or global circulation models to search for the "ghost" of this behavior in the full complex dynamics with many degrees of freedom [**39, 40, 24**]. In this context, explicit stochastic models involving Markov jump processes have been utilized to mimic random jumps between different basins of low frequency variability in observations [**45, 27**]. Also, the effects of small random additive noise have been analyzed in a similar fashion [**12**] when introduced in an ad-hoc fashion in the simplest scenario for multiple equilibria [**6**].

The work mentioned in the above paragraph requires regression fitting of an assumed linear stochastic model to the observed climate and its fluctuations. By necessity this approach to stochastic modeling has severe limitations for predicting climate change and diagnosing its causes. Alternative systematic a priori stochastic mode reduction strategies are needed for more practical prediction and improved regression fitting. This is the main focus of the present chapter, which is based on a systematic mathematical strategy for stochastic mode reduction developed recently by Majda, Timofeyev, and Vanden–Eijnden(MTV) in a series of papers [**34, 35, 36, 37**]. The stochastic mode reduction technique is a two step procedure based on the assumption that the degrees of freedom under consideration have been split into essential resolved modes whose dynamics is the main objective and unessential unresolved modes which are to be eliminated in the reduced description. In the first step of the procedure, the equations of motion for the unresolved modes are modified by representing the nonlinear self-interactions terms between unresolved modes by stochastic terms. The motivation is that the self-interaction terms are responsible for the sensitive dependence on small perturbations in the system on short time scales, and that these modes can indeed be represented by stochastic terms if coarse-grained modeling on longer time scales, as for climate, is the objective. This step in the procedure is called *stochastic consistency*. In the second step of the procedure, the equations of motion for the unresolved modes are then eliminated using standard projection technique for stochastic differential equations [**26, 29, 19, 34, 35**]. The elimination step is mathematically rigorous in the limit where the stochastic terms are infinitely fast, corresponding to the situation where the unresolved modes evolve much faster than the resolved ones. This step is called *stochastic mode elimination.*

This systematic mode reduction technique has two obvious advantages. First, the ad-hoc simplification of the original dynamics is made on the level of the unresolved modes rather than the resolved ones. This is unlike the procedures described in the second paragraph above which are in common use in the atmospheric science community where the approximations are developed on the resolved modes, one drops all couplings with the unresolved modes, and replaces them by ad-hoc stochastic terms of linear Langevin type, with parameters obtained by regression fitting. In contrast, the systematic mode reduction procedure predicts the structure of the stochastic terms in the equations for the resolved modes [**34, 35**] and this structure can be surprisingly rich: nonlinear corrections arise, multiplicative noises occur, both stable and unstable linear Langevin terms arise, or all in combination, depending on the quantitative features of the underlying dynamics [**35**]. The second advantage of the technique is its rigor in an appropriate limiting parameters range which can be deduced *a priori* from the original equations. This provides a guideline for the applicability of the method which is an important feature of the procedure.

The MTV procedure has been applied to a barotropic climate model on a sphere with realistic orography by Franzke, et al. [**17**]. In that paper, the MTV procedure reduces the dynamics of a climate model with roughly a thousand variables to a systematic low order stochastic model with two, four, or seven degrees of freedom derived from the MTV procedure; the reduced model predicts a priori comparable mean climate, variances, turbulent eddy fluxes, correlation functions as in the many degree of freedom climate model. At the present time other reduced order methods based on a more *ad hoc* closure strategies seem to require at least fifty modes to even begin to achieve comparable results [**31**]. In another direction, Egger [**14**] has utilized the systematic strategies of MTV for modeling angular momentum [**34, 37**] to develop new regression strategies for analyzing observational data. Additional applications of the MTV procedure are currently underway for even more realistic climate models in works of Franzke and the first author.

How have such applied practical results been achieved for such a mathematical strategy? New phenomena such as augmented nonlinearity and multiplicative noise arise in the approach. Very subtle issues occur in practical systems because the mathematical theory is strictly valid only when a certain parameter, ε, measuring the ratio of correlation times is small, $\varepsilon \ll 1$, yet in the practical applications, ε sometimes has large values like $\varepsilon \simeq 0.5$, 0.7. Also, practical systems have some aspects of statistical and stochastic variability which are difficult to measure and estimate. In the MTV work, all of these issues are addressed through the design of systematic families of simplified models ranging from three mode systems satisfying basic physical symmetries to such models coupled to a turbulent bath of unresolved modes given by the TBH model discussed earlier in Chapter 1 and Section 2.6 which has an a priori predicted spectrum and correlation scaling theory which is well understood through mathematical reasoning. The key feature is to design a hierarchy of models so that the simplest models are essentially exactly solvable with unambiguous mathematical theory that can be checked for any value of ε including both $\varepsilon \ll 1$ and $\varepsilon = O(1)$ while the more complex models have simple unambiguous behavior predicted from mathematical theory even though they are not exactly solvable. This approach has been utilized throughout the MTV research program and in this fashion, the new predicted phenomena have been shown to be robust and significant; furthermore, confidence has been gained in the procedure as a practical strategy even when $\varepsilon = O(1)$ under appropriate circumstances.

Another use of the development of simpler models in the MTV work is to elucidate the strengths and weaknesses of the completely linear stochastic regression fitting strategies described above in the second paragraph with the alternative approach involving deterministic bifurcation theory for reduced climate variables mentioned in the third paragraph above. Intrinsic stochastic dynamics arises in deterministic systems with many degrees of freedom through the backscatter of rapidly varying chaotic dynamics onto slowly varying climate modes; this is discussed briefly in a simple model in Section 3.2 below. Thus, the above issues can be rephrased as the competition between intrinsic stochastic noise and projected nonlinear dynamics in complex models with many degrees of freedom. The simplified models developed in the MTV approach have been used for elucidating the role of this competition in systems with stable periodic orbits and multiple equilibria [**37**] and recently, to elucidate the competition between heteroclinic chaos and intrinsic stochastic noise [**33**] in systems with many degrees of freedom.

In Section 3.3, the MTV strategy is briefly summarized. In Section 3.4 we illustrate the approach described above through the development of a new simplified model, where a number of the above issues can be addressed unambiguously with firm mathematical underpinnings; these are examples of elementary models with a sound physical basis yet with surprisingly complex features emerging. Much additional work remains to be done on reduced stochastic modeling for complex systems; the authors hope that this chapter entices others to examine this important and fascinating topic.

3.2. Intrinsic stochastic noise in deterministic dynamical systems

The simplest stochastic model for a variable, x, is given by the Langevin equation,

$$dx = -\gamma x \, dt + \sigma \, dW(t) \tag{3.1}$$

where $W(t)$ is the Wiener process so that $dW(t)$ is white noise [**2, 19**]. In his pioneering paper on stochastic climate modeling, Hasselmann [**23**] considered the mid-latitude atmospheric surface fluxes, which are rapidly varying in time compared with ocean dynamics, as white noise forcing, $\sigma \, dW(t)$ in (3.1), on the mixed surface layer of the ocean represented by the variable x. Hasselmann showed that this white noise forcing produced a significant red spectrum response in the ocean through the exact solution of (3.1).

Next, consider the deterministic time discrete models with the form

$$\begin{aligned} y_{n+1} &= \Phi(y_n), \\ x_{n+1} &= \lambda \, x_n + y_{n+1}, \end{aligned} \tag{3.2}$$

with a fixed value of λ, $0 < \lambda < 1$. Formally, we identify the parameter λ in (3.2) with the parameter, γ, in (3.1) through $\lambda = \exp(-\gamma \Delta t)$ with Δt the time unit. If $\Phi(y)$ is one of the well-known strongly mixing deterministic ergodic maps of the unit interval, the system in (3.2) is known as the Kaplan–Yorke map; note that the variable x is forced by y at each time step. The simplest version of these models occurs when $\Phi(y_n)$ is a map of the unit interval, $0 \le y \le 1$ into itself given by the Bernoulli shift,

$$\Phi(y_n) = 2y_n \mod 1. \tag{3.3}$$

The models in (3.3) provide explicit examples of chaotic dynamical systems with complex (fractal) features changing as λ is varied [**3**]. Now the system in (3.2) has chaotic dynamics but is completely deterministic while the equation in (3.1) is stochastic. Nevertheless, it can be proved with full mathematical rigor that if $\lambda \nearrow 1$ in a suitable fashion, the x-variable of the solution of the deterministic chaotic dynamical system in (3.2) converges to the solution of the fully stochastic Langevin equation in (3.1). A basic reference for the chaotic dynamics of (3.2) and this result is the book by Beck and Schlögl [**3**]. This is the simplest example, where a deterministic chaotic dynamical system with chaotic dynamics on a short time scale acquires a completely stochastic character for some of the slower variables, x, in this case. Returning to Hasselmanns example, in the model in (3.2) one can think of the variables y as the faster atmospheric dynamics and the variable x as the slower ocean dynamics. The mathematical theory guarantees that in a suitable regime, the behavior of the coarse-grained climate variable, x, instead can be studied through the simpler stochastic equation (3.1) for the variable x, alone. It is useful for the reader to remember this elementary example of intrinsic stochastic noise when we discuss the MTV procedure in the next section.

3.3. The MTV strategy for stochastic mode reduction

The dynamical core of comprehensive general circulation models (GCM) for the atmosphere or ocean is governed essentially by quadratically nonlinear equations for a variable $\vec{z}$ belonging to $\mathbb{R}^N$ with $N \gg 1$ with the general form [**38, 20**],

$$\frac{d\vec{z}}{dt} = [L + D]\vec{z} + B(\vec{z}, \vec{z}) + \vec{F}. \tag{3.4}$$

In GCMs, the dimension N is often several billion dimensions and roughly of order 10,000 for the dynamical core away from boundaries. As mentioned in the introduction to this chapter, there is a great practical interest in deriving reduced models with many fewer degrees of freedom for systems like those in (3.4) which can capture the important long-time low frequency variability of such a system with reasonable accuracy. As discussed in the introduction, the MTV procedure is a systematic mathematical strategy which attempts to achieve this. We discuss this strategy briefly below in this section but first we discuss the general structure of the system in (3.4).

The linear operator, $L + D$, in (3.4) has two different contributions [**38, 20**].

A) L, a skew-symmetric operator representing the β-effect of Earth's curvature, topography, etc. and satisfying,

$$L^\top = -L. \tag{3.5}$$

B) D, a negative definite symmetric operator,

$$D^\top = D, \tag{3.6}$$

representing dissipative processes such as surface drag, radiative damping, viscosity, etc.

The quadratic operator $B(\vec{z}, \vec{z})$ conserves the energy $E = \frac{1}{2}\vec{z} \cdot \vec{z}$ by itself so that necessarily B satisfies

$$\vec{z} \cdot B(\vec{z}, \vec{z}) = 0, \tag{3.7}$$

where $\vec{z} \cdot \vec{w}$ denotes the Euclidean inner product for simplicity here. Finally $\vec{F}$ represents the effects of external solar forcing, etc.

In the MTV procedure [**34, 35, 36, 37**] the variable $\vec{z}$ is decomposed into an orthogonal decomposition through the variables $\vec{x}$, $\vec{y}$, where $\vec{z} = [\vec{x}, \vec{y}]$. The variable $\vec{x}$ denotes the resolved climate variables of the system which necessarily, as climate variables, evolve slowly in time compared with the $\vec{y}$ variables which evolve more rapidly in time. For example in the atmosphere, $\vec{x}$, might represent a small set of the low frequency planetary scale teleconnection patterns such as the Pacific North American pattern [**17**] while $\vec{y}$ represents the other variables. The MTV-strategy begins with an exact projection of the dynamics in (3.4) into its $\vec{x}$ and $\vec{y}$ components,

$$\begin{aligned}\frac{d\vec{x}}{dt} &= \big([L_{11} + D_{11}]\vec{x} + B^1_{11}(\vec{x}, \vec{x})]\big) + [L_{12} + D_{12}]\vec{y} + B^1_{12}(\vec{x}, \vec{y}) \\ &\qquad + B^1_{22}(\vec{y}, \vec{y}) + F_1 \\ \frac{d\vec{y}}{dt} &= [L_{22} + D_{22}]\vec{y} + [L_{21} + D_{21}]\vec{x} + F_2 + B^2_{11}(\vec{x}, \vec{x}) + B^2_{12}(\vec{x}, \vec{y}) \\ &\qquad + B^2_{22}(\vec{y}, \vec{y}).\end{aligned} \tag{3.8}$$

In the first step of the MTV procedure, the nonlinear self-interactions of the fast variables by themselves, $B^2_{22}(\vec{y},\vec{y})$ in (3.8), are replaced by damping and stochastic forcing as an *approximation*

$$B^2_{22}(\vec{y},\vec{y}) \simeq -\frac{\Gamma}{\varepsilon}\vec{y} + \frac{\sigma}{\sqrt{\varepsilon}}\dot{\vec{W}}, \tag{3.9}$$

where Γ and σ are symmetric matrices (usually chosen diagonal) with Γ negative definite. The motivation for (3.9) is that the nonlinear self-interacting terms, $B^2_{22}(\vec{y},\vec{y})$, of the fast variables are responsible for the chaotic sensitive dependence on small perturbations in (3.4) and do not require a more detailed description if their effect on the coarse-grained dynamics for the climate variables alone is the main objective. It is also useful for the reader to recall the simple models with intrinsic stochastic noise from Section 3.2 above as providing some elementary motivation. Note that the crucial nonlinear coupling between the $\vec{x}$ and $\vec{y}$ variables is retained with complete fidelity when the approximation from (3.9) is utilized in (3.8). In practical applications, the variables $\vec{y}$ are associated with turbulent cascade dynamics and intuitively, their detailed self-interaction through $B^2_{22}(\vec{y},\vec{y})$ is not needed if a reduced stochastic description of the slow dynamics is required. The parameter ε in (3.9) measures the time rate of the nonlinear self-interaction processes of the unresolved variables $\vec{y}$ compared to the rates of the nonlinear interactions in the system. The approximation in (3.9) is an assumption and there is no rigorous mathematical theory which justifies it directly. Once the approximation from (3.9) is inserted in (3.8) a stochastic system of ordinary differential equations (ODEs) results rather than the original deterministic system.

In the second step of the MTV procedure, it is assumed that $\varepsilon \ll 1$ so that standard techniques of averaging and adiabatic elimination of fast modes in stochastic equations [**26, 29, 19**] can be applied. This technique yields a new stochastic equation for the unresolved climate variables, $\vec{x}$ alone on a longer (climate) time scale which solve the *Reduced Stochastic Equations*,

$$d\vec{x} = \vec{F}_{\text{eff}}(\vec{x}) + \vec{G}_{\text{eff}}(\vec{x})\,d\vec{W}, \tag{3.10}$$

where the coefficients $\vec{F}_{\text{eff}}(\vec{x})$ and $\vec{G}_{\text{eff}}(\vec{x})$ are determined explicitly by formulas through the explicit interaction coefficients in (3.8), (3.9) [**34, 35**]. A very instructive elementary example of this procedure is developed in the remaining sections of this chapter, which shows the new nonlinear phenomena which emerge through this procedure. Clearly the net effect of the two steps in the MTV procedure is a *stochastic mode reduction* where the deterministic system in (3.4) has been replaced by the much lower dimensional set of stochastic equations in (3.10) evolving on the longer time scale. Confidence in the above two-step procedure has been gained by testing it on a wide variety of deterministic models with hundreds of degrees of freedom both for $\varepsilon \ll 1$ and $\varepsilon = O(1)$ where the reduced variables range over a small dimension ranging from two to seven dimensions in various applications [**34, 36, 37, 33, 17**]. In practical applications, one needs a rapid way to compute the explicit equations in (3.10) without directly fitting the matrices, Γ, σ in (3.9) to the observed correlations and variances; such a seamless MTV method is developed in Franzke et al. [**17**] and reinterpreted through the method of averaging.

One important practical issue not addressed by the MTV procedure is the decomposition of the state variable $\vec{z} = (\vec{x},\vec{y})$ into resolved slowly varying climate variables, $\vec{x}$, and unresolved variables, $\vec{y}$. The variables $\vec{x}$ can be determined in a variety of ways such as empirical orthogonal functions, optimal persistence patterns, principal interaction patterns, etc. [**9, 30, 31, 8**]; it is evident that the choice of the optimal basis for determining the climate variables, $\vec{x}$, can significantly affect the performance of the above stochastic mode reduction strategy. This area is a major future research topic for stochastic mode reduction strategies. The paper of Crommelin and Majda [**8**] tests the variety of optimal basis strategies on a low order atmospheric model with low frequency regime transitions and is a good starting point for the interested reader.

3.4. A simple model for the competition between intrinsic stochastic noise and deterministic dynamics

The models which we introduce here are the simplest version of the general geophysical models in (3.4) with the structure in (3.6), (3.7) where $\vec{z} = (\vec{x}, \vec{y})$ is three-dimensional with two climate variables, $\vec{x} = (x_1, x_2)^\top$ and a single unresolved variable, y with the stochastic terms in (3.9) added to y to represent the nonlinear interactions of y with other unresolved modes. We write this model below in the projected form from (3.8) for the variables $\vec{x} = (x_1, x_2)^\top, y$. First, a general quadratic nonlinear operator that satisfies (3.7) in order to guarantee the energy conservation principle is given by

$$
\begin{aligned}
\frac{d\vec{x}}{dt} &= a_1 \vec{x}^\perp x_1 + a_2 \vec{x}^\perp x_2 + \begin{bmatrix} b_1 x_2 y \\ b_2 x_1 y \end{bmatrix}, \\
\frac{dy}{dt} &= b_3 x_1 x_2,
\end{aligned} \tag{3.11}
$$

provided the interaction coefficients satisfy

$$
b_1 + b_2 + b_3 = 0. \tag{3.12}
$$

In (3.11), $\vec{x}^\perp = [-x_2, x_1]^\top$ and a_1, a_2 are arbitrary coefficients. The reader can readily verify that this system conserves the energy, $E = \frac{1}{2}(\vec{x}\cdot\vec{x} + y^2)$. The complete model simply adds linear terms with the structure in (3.5), constant forcing terms, and the stochastic forcing in (3.9) to y. The result is the following model which we study in detail in the subsequent sections of this chapter:

$$
\begin{aligned}
\frac{d\vec{x}}{dt} &= D\vec{x} + (\omega + a_1 x_1 + a_2 x_2)\vec{x}^\perp + \vec{F} + \vec{L} y + \begin{bmatrix} b_1 x_2 y \\ b_2 x_1 y \end{bmatrix}, \\
\frac{dy}{dt} &= -\vec{L}\cdot\vec{x} + b_3 x_1 x_2 + F_3 - \frac{\gamma}{\varepsilon} y + \frac{\sigma}{\sqrt{\varepsilon}} \dot{W}(t).
\end{aligned} \tag{3.13}
$$

In (3.13), we have assumed that the dissipation matrix D is diagonal, also $\vec{L}$ is a fixed vector with the form in (3.13) guaranteeing skew-symmetry, while ω is a constant rotation frequency. Without loss of generality, below we assume that the forcing $\vec{F}$ has the form $\vec{F} = [F, 0]^\top$.

The elementary model in (3.13) has all the physical structural conditions of (3.4) with the explicit effects of interaction with unresolved features modeled by stochastic forcing. It has the additional advantage that for $\varepsilon \ll 1$, the reduced stochastic equations (3.10) can be derived explicitly by the direct method; see [**34, 35, 36, 37**] for simple examples of this sort. Furthermore, in the next section we show that the derived effective coefficients in $\vec{F}_{\text{eff}}(\vec{x})$ and $\vec{G}_{\text{eff}}(\vec{x})$ exhibit a remarkable combination of competing effects of augmented nonlinearity, additive, and multiplicative stochastic noises which have clear physical origins. Furthermore, the predictions of the MTV stochastic mode reduction procedure for large values of $\varepsilon = O(1)$ can be checked with the behavior of the complete stochastic system in (3.13) through accurate controlled numerical experiments. In addition, the nonlinear deterministic dynamics for climate variables, $\vec{x}$, in the model in (3.13) is given by the ODEs,

$$\begin{aligned} \frac{d\vec{x}}{dt} &= D\vec{x} + (\omega + a_1x_1 + a_2x_2)\vec{x}^{\perp} + \vec{F} \\ &\equiv \vec{F}_0(\vec{x}; F). \end{aligned} \tag{3.14}$$

As shown in Section 3.6 below, as we vary the climatological forcing, F, as a bifurcation parameter, the ODEs in (3.14) can have a unique steady state, multiple steady states, homoclinic trajectories, and unstable periodic orbits for various parameter regimes which are readily understood through phase plane analysis. With all the above features, the models in (3.13) are elementary yet very instructive examples for the central problem described in the introduction involving the competition between projected nonlinear dynamics and intrinsic stochastic noise. This is the topic of Section 3.7 below.

3.5. Explicit stochastic mode reduction in the model

We now apply the MTV procedure to the model in (3.13) to obtain explicit reduced stochastic equations for the slow variables, $\vec{x}$. Because the model is elementary we can use the direct method developed in [**34, 35, 36, 37**], without introducing a longer coarse-grained time scale as utilized in the general procedure. Further details on the derivation below also can be found in these references.

The stochastic differential equation (SDE) for the variable y in (3.13) is linear in y. Thus, given $\vec{x}(t)$ its solution is [**2**]

$$y(t) = e^{-\gamma t/\varepsilon}y(0) + \int_0^t e^{-\gamma(t-s)/\varepsilon}[b_3x_1(s)x_2(s) - \vec{L}\cdot\vec{x}(s) + F_3(s)]\,ds + g(t), \tag{3.15}$$

where

$$g(t) = \frac{\sigma}{\sqrt{\varepsilon}}\int_0^t e^{-\gamma(t-s)/\varepsilon}\,dW(s). \tag{3.16}$$

Inserting (3.15) into the first two equations in (3.13) for the variable $\vec{x}$ yields an exact, non-Markovian system of equations for $\vec{x}(t)$.

Since we are interested in the long-time statistical behavior of the climate variables $\vec{x}(t)$ as $\varepsilon \to 0$, we consider the limit as $\varepsilon \to 0$ of the three terms on the right of (3.15). First, we immediately have

$$e^{-\gamma t/\varepsilon}y(0) \to 0. \tag{3.17}$$

Second, using integration by parts we find

$$\begin{aligned} &\int_0^t e^{-\gamma(t-s)/\varepsilon}[b_3x_1(s)x_2(s) - \vec{L}\cdot\vec{x}(s) + F_3]\,ds \\ &\qquad \to \frac{\varepsilon}{\gamma}[b_3x_1(t)x_2(t) - \vec{L}\cdot\vec{x}(t) + F_3]. \end{aligned} \tag{3.18}$$

Finally, it can be shown that $g(t)$ is itself approximatively white-noise as $\varepsilon \to 0$, i.e.,

$$g(t)\,dt \to \sqrt{\varepsilon}\frac{\sigma}{\gamma}\,dW(t). \tag{3.19}$$

We note, however, that, as an approximation of a process with finite correlation time, $dW(t)$ has to be interpreted in the Stratonovich sense [**2**].

Combining these formulas in the first two equations of (3.13), we obtain the following Stratonovich system of SDEs:

$$\begin{aligned} d\vec{x}(t) &= \vec{F}_0(\vec{x};F)\,dt + \frac{\varepsilon}{\gamma}\vec{L}[b_3x_1x_2 - \vec{L}\cdot\vec{x} + F_3]\,dt \\ &+ \frac{\varepsilon}{\gamma}\begin{bmatrix} b_1x_2(t)\{b_3x_1(t)x_2(t) - \vec{L}\cdot\vec{x}(t) + F_3\} \\ b_2x_1(t)\{b_3x_1(t)x_2(t) - \vec{L}\cdot\vec{x}(t) + F_3\}\end{bmatrix} dt \\ &+ \sqrt{\varepsilon}\frac{\sigma}{\gamma}\begin{bmatrix} L_1 + b_1x_2(t) \\ L_2 + b_2x_1(t)\end{bmatrix} \circ dW(t). \end{aligned} \tag{3.20}$$

This system is equivalent to the Itô system [**2, 19**]

$$\begin{aligned} d\vec{x}(t) &= \vec{F}_0(\vec{x};F)\,dt + \frac{\varepsilon}{\gamma}\vec{L}[b_3x_1x_2 - \vec{L}\cdot\vec{x} + F_3]\,dt \\ &+ \frac{\varepsilon}{\gamma}\begin{bmatrix} b_1x_2(t)\{b_3x_1(t)x_2(t) - \vec{L}\cdot\vec{x}(t) + F_3\} \\ b_2x_1(t)\{b_3x_1(t)x_2(t) - \vec{L}\cdot\vec{x}(t) + F_3\}\end{bmatrix} dt \\ &+ \varepsilon\frac{\sigma^2}{2\gamma^2}\left(b_1b_2\vec{x}(t) + \begin{bmatrix} b_1L_2 \\ b_2L_1\end{bmatrix}\right) dt + \sqrt{\varepsilon}\frac{\sigma}{\gamma}\begin{bmatrix} L_1 + b_1x_2(t) \\ L_2 + b_2x_1\end{bmatrix} dW(t). \end{aligned} \tag{3.21}$$

It yields reduced stochastic equations as in (3.10) for the slow climate variable, $\vec{x}$, with

$$\vec{F}_{\text{eff}}(\vec{x}) = \vec{F}_0(\vec{x};F) + \varepsilon\vec{F}'(\vec{x}). \tag{3.22}$$

In addition to the deterministic dynamics term, $\vec{F}_0$, the effective forcing now contains remarkable new features in $\vec{F}'$, such as augmented quadratic and cubic nonlinearity, altered mean forcing, and cross-diagonal dissipation, while the effective noise coefficient, $\vec{G}_{\text{eff}}$, contains both additive and multiplicative stochastic noise. Recall that the parameter ε measures the time correlation between the slow variables, $\vec{x}$, and the fast variable, y. Although the derivation above is only rigorous in the asymptotic limit, $\varepsilon \to 0$, we will see in Section 3.7 below, that even when $\varepsilon = O(1)$, the reduced stochastic equations in (3.21) typically represent with high fidelity the long-term statistical behavior of the original system (3.13). We further remark that for moderately large values of ε, both the augmented nonlinearity and added dissipation can directly alter the qualitative behavior of the deterministic dynamics, which is no longer dictated by $\vec{F}_0(\vec{x};F)$ only.

To better understand the meaning of the various new terms in (3.21), we first consider two special cases, which emphasize their individual contributions, and subsequently discuss their combined effect.

Case 1. Linear additive coupling. To begin we consider (3.13) with $b_1 = b_2 = b_3 = 0$. Then the new terms that appear in the reduced dynamics equations (3.21) consist of

$$\vec{F}'(\vec{x}) = \frac{1}{\gamma}\vec{L}F_3 - \frac{1}{\gamma}\vec{L}\vec{L}^\top \vec{x} \tag{3.23}$$

$$\vec{G}_{\text{eff}}(\vec{x}) = \sqrt{\varepsilon}\,\frac{\sigma}{\gamma}\vec{L}. \tag{3.24}$$

They involve augmented linearity and a new effective forcing term. Because the matrix $\vec{L}\vec{L}^\top$ is symmetric positive semi-definite, the augmented linear term is still damping; however, the dissipation is no longer purely diagonal, in contrast to the dissipation matrix D of the deterministic dynamics in (3.13). The small scale forcing term, F_3, induces a significant shift in the forcing modulated by $\vec{L}$.

To illustrate the effect of induced cross-diagonal damping, we choose a trivial scenario with $\vec{F}_0 \equiv 0$, $F_3 = 0$, $\vec{L} = [-\frac{1}{2}, \frac{1}{2}]^\top$, $\gamma = \sigma = 1$, and set $\varepsilon = 0.5$. Hence the

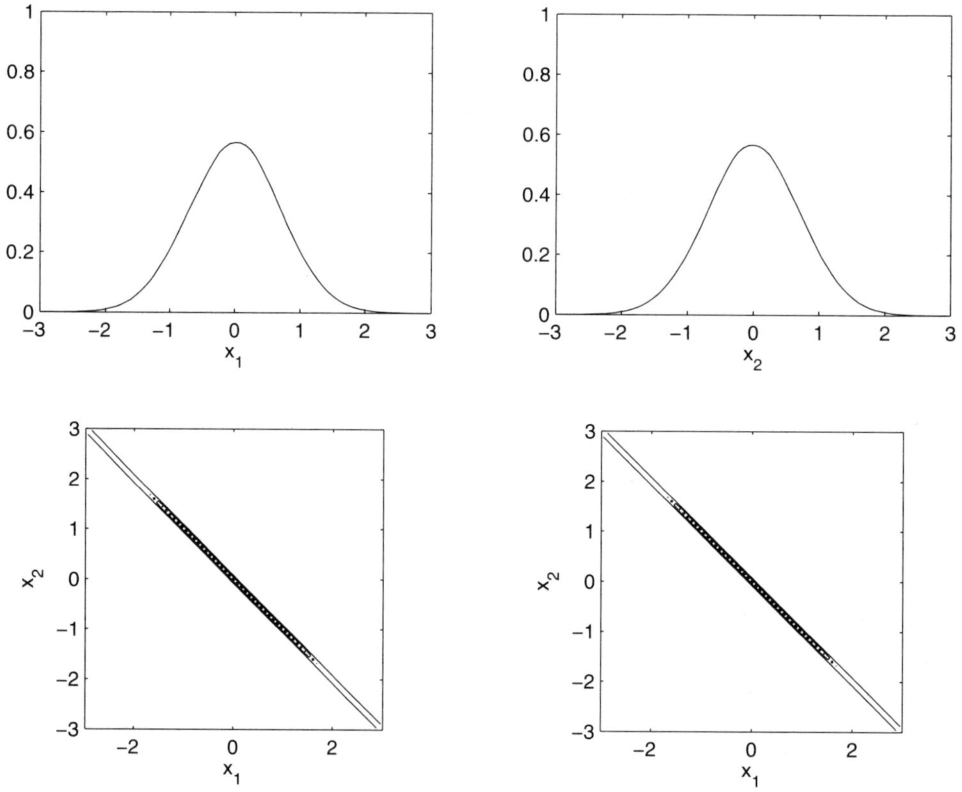

FIGURE 3.1. Linear additive coupling: comparison of the marginal and joint PDFs of x_1 and x_2 with $\vec{L} = [-\frac{1}{2}, \frac{1}{2}]^\top$, $\vec{F}_0 = 0$, $F_3 = 0$, and $\varepsilon = 0.5$. Top: the marginal PDF for the stochastic dynamics (plain) and for the reduced stochastic equations (dashed) are shown—here they coincide. Bottom: the joint PDF for the stochastic dynamics (left) and the reduced dynamics (right). Note the purely one-dimensional nature of the induced cross-diagonal diffusion.

solution of the deterministic dynamics is constant and reduces to $\vec{x}(t) \equiv \vec{x}_0$, with $\vec{x}_0 = [0, 0]^\top$. In contrast, the solution $\vec{x}(t)$ of (3.21) is now a stationary Gaussian process with mean $\vec{x}_0$, which only varies along the direction $\vec{L}$ in the (x_1, x_2)-plane—note that $\vec{L}$ is an eigenvector of $\vec{L}\vec{L}^\top$ associated with the eigenvalue $\frac{1}{2}$. In Figure 3.1 contours of the joint probability distribution functions (PDF) for x_1 and x_2 are shown, both for the full stochastic system (3.13) and for the reduced dynamics equations (3.21). Both essentially coincide despite the rather large value for $\varepsilon = 0.5$. The fast stochastic variable y induces a cross-diagonal, one-dimensional diffusion along the direction $\vec{L}$, which is perfectly replicated by the reduced model from the MTV procedure.

Case 2. Nonlinear triad interactions. Next, we consider (3.13) with $\vec{L} = 0$. In that case, the new terms that appear in (3.21) reduce to

$$\vec{F}'(\vec{x}) = \frac{1}{\gamma}\left(\begin{bmatrix} b_1 x_2(t)\{b_3 x_1(t) x_2(t) + F_3\} \\ b_2 x_1(t)\{b_3 x_1(t) x_2(t) + F_3\}\end{bmatrix} + \frac{\sigma^2}{2\gamma} b_1 b_2 \vec{x}\right) \tag{3.25}$$

$$\vec{G}_{\text{eff}}(\vec{x}) = \sqrt{\varepsilon}\,\frac{\sigma}{\gamma}\begin{bmatrix} b_1 x_2 \\ b_2 x_1 \end{bmatrix}. \tag{3.26}$$

Here we find both augmented linearity and augmented cubic nonlinearity. Unless $b_1 = -b_2$, the added nonlinear terms will not be skew-symmetric, and hence affect the energy budget, depending on b_1, b_2, and b_3. Note that this is completely natural because the nonlinear terms from (3.11), (3.12) conserve the total energy but, in general, allow energy exchange between $\vec{x}$ and y. In addition we find a (standard) "Itô drift" term in (3.25), which results from the previous analysis. These new terms and their effects in the absence of underlying deterministic dynamics, that is with $\vec{F}_0 \equiv 0$, are studied in detail in [**36**]. Note that on the coarse-grained longer time scale, T, with $T = \varepsilon t$, the factors of ε from (3.22), (3.26) disappear to recover the form in (3.11).

Case 3. General set-up. Finally, we consider the full reduced stochastic equations given by (3.13). Here we rewrite (3.13) in a slightly different form to emphasize the additional terms that result from the combined effect of additive coupling and nonlinear triad interactions. Then we have

$$\begin{aligned} \vec{F}'(\vec{x}) &= \frac{1}{\gamma}(\vec{L}F_3 - \vec{L}\vec{L}^\top\vec{x}+) \\ &\quad + \frac{1}{\gamma}\left(\begin{bmatrix} b_1 x_2(t)\{b_3 x_1(t) x_2(t) + F_3\} \\ b_2 x_1(t)\{b_3 x_1(t) x_2(t) + F_3\}\end{bmatrix} + \frac{\sigma^2}{2\gamma} b_1 b_2 \vec{x}\right) \\ &\quad + \frac{1}{\gamma}\left(b_3 \vec{L} x_1 x_2 - \vec{L}\cdot\vec{x}\begin{bmatrix} b_1 x_2 \\ b_2 x_1\end{bmatrix} + \frac{\sigma^2}{2\gamma}\begin{bmatrix} b_1 L_2 \\ b_2 L_1\end{bmatrix}\right) \end{aligned} \tag{3.27}$$

$$\vec{G}_{\text{eff}}(\vec{x}) = \sqrt{\varepsilon}\frac{\sigma}{\gamma}\begin{bmatrix} L_1 + b_1 x_2 \\ L_2 + b_2 x_1 \end{bmatrix}. \tag{3.28}$$

Besides the terms, which appeared in the previous two cases, the new coupling terms in $\vec{F}'_{\text{eff}}$ appear last in (3.27). They involve two terms with augmented quadratic nonlinearity in $\vec{F}'_{\text{eff}}$ and one additional "Itô drift" term.

3.6. Bifurcation theory in the projected deterministic dynamics

Here we choose specific values for the parameters in (3.14) and analyze in detail the bifurcation sequence that the purely deterministic dynamics experience as we

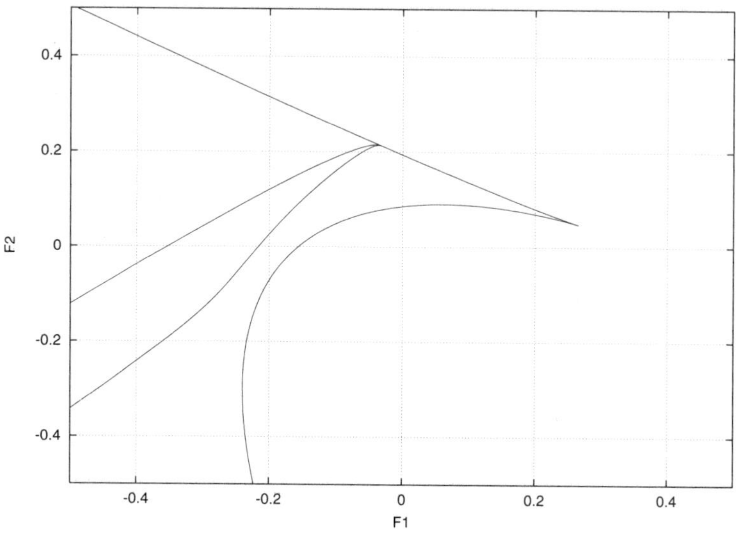

FIGURE 3.2. The bifurcation diagram of (3.14) with the parameter settings in (3.29) is shown for varying $\vec{F} = (F_1, F_2)$.

vary the climatological forcing. The same parameter values will then be used in the numerical experiments for the fully stochastic regime in Section 3.7.

The nonlinear deterministic dynamics for the climate variables, $\vec{x}$, in the model in (3.13) is dictated by (3.14). Hence, we now fix the parameter values D, a_1, a_2, and ω in (3.14) and discuss the behavior of the resulting dynamical system as we vary the climatological forcing $\vec{F} = [F, 0]^\top$. Thus, we choose the parameter values

$$D = \frac{1}{10}\begin{bmatrix} -2 & 0 \\ 0 & -1 \end{bmatrix}, \quad a_1 = 1, \; a_2 = -1, \; \omega = 1. \tag{3.29}$$

Then (3.14) reduces to

$$\begin{aligned} \frac{dx_1}{dt} &= -x_2(1 + x_1 - x_2) - \frac{2}{10}x_1 + F_1, \\ \frac{dx_2}{dt} &= x_1(1 + x_1 - x_2) - \frac{1}{10}x_2. \end{aligned} \tag{3.30}$$

In the absence of forcing and dissipation, that is when $F = 0$ and $D = 0$, the dynamical system has an isolated equilibrium at $x_1 = x_2 = 0$, and a continuous set of equilibria located on the straight line $\Omega = \{(x_1, x_2) \mid 1+x_1-x_2 = 0\}$. With added forcing and dissipation the infinite set of equilibria on Ω vanishes; nevertheless, its signature will remain apparent in the phase portraits below. Note that the vector field on the right hand side of (3.30) points inward on the circle, $|\vec{x}| = R$, for R large enough; hence, all solutions of (3.30) stay bounded.

As we vary the climatological forcing, F, as a bifurcation parameter, the behavior of the dynamical system described by (3.30) qualitatively changes. The full bifurcation diagram[1] is shown in Figure 3.2 for the more general constant forcing $\vec{F} = [F_1, F_2]^\top$. Since $\vec{F} = [F, 0]^\top$ in (3.30), we now discuss in detail the different

[1]generated by D. Crommelin with the software AUTO [**13**]

TABLE 3.1. Locations of *stable* equilibria for varying climatological forcing $\vec{F} = [0, F]$.

		Lower equilibrium		Higher equilibrium	
$F = -0.1$	low state	$x_1 = -0.01$	$x_2 = -0.09$		
$F = -0.18$	competing states	$x_1 = -0.01$	$x_2 = -0.16$	$x_1 = -0.49$	$x_2 = 0.63$
$F = -0.25$	competing states	$x_1 = -0.02$	$x_2 = -0.21$	$x_1 = -0.39$	$x_2 = 0.82$
$F = -0.5$	high state			$x_1 = -0.28$	$x_2 = 1.12$

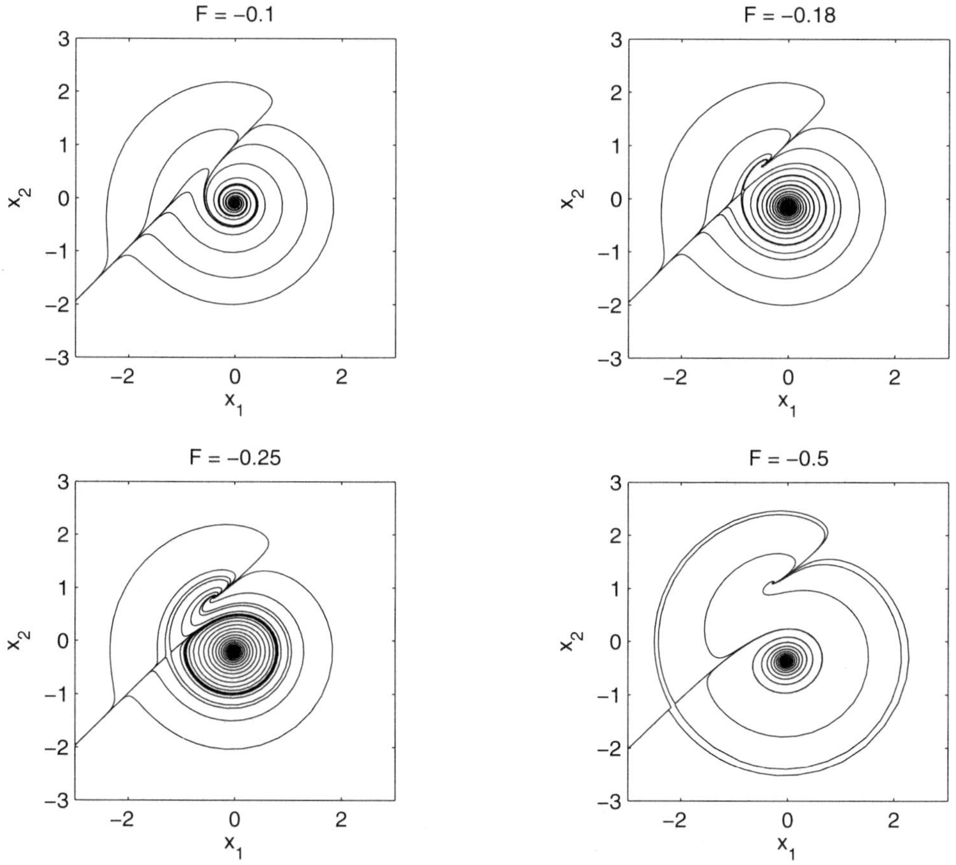

FIGURE 3.3. Phase portraits of the purely deterministic dynamics (3.30) for different values of the bifurcation parameter $F = -0.1$, -0.18, -0.25, -0.5.

regimes that occur, as F varies along the line $F_2 = 0$ in Figure 3.2. Three bifurcations occur on that line at the intersection points $\mu_1 = -0.15631$, $\mu_2 = -0.21639$, and $\mu_3 = -0.35336$. The four intervals, $F \leq \mu_3$, $\mu_3 \leq \mu_2$, $\mu_2 \leq \mu_1$, and $\mu_1 \leq F$ correspond to qualitatively distinct regimes. Inside each interval we choose a representative value for F: $F = -0.5, -0.25, -0.18, -0.1$.

For $F = -0.1$ the dynamical system has a unique steady state close to the origin, whose coordinates are given in Table 3.1. The signature of Ω splits the (x_1, x_2)-plane in half, as shown in Figure 3.3, while all solutions spiral towards the equilibrium, a "low" climate state because of its small energy content. Moving to the left along the F_1-axis, a saddle-node bifurcation occurs as $F = F_1$ crosses the value μ_1. Beyond μ_1, two new equilibria appear in the vicinity of Ω, one stable and the other unstable. We denote by "high" climate state that new, more energetic, stable equilibrium. As shown in Figure 3.3, for $F = -0.18$ both stable equilibria, the high and the low climate states, now compete as trajectories eventually reach either one of them; their coordinates are given in Table 3.1. We remark that the basin of attraction of the low state, albeit unbounded, is smaller than that of the high state. As F crosses the next bifurcation point at μ_2, a homoclinic orbit appears which connects the unstable equilibrium to itself, isolating the basin of attraction of the low state from the surrounding unbounded exterior region. Past μ_2, the homoclinic orbit transforms into a repelling periodic orbit, effectively dividing the plane in two competing basins of attraction: inside, all trajectories slowly spiral clockwise towards the low climate state whereas outside, all trajectories rapidly reach the high climate state. This is illustrated for the value, $F = -0.25$, in Figure 3.3. Finally, as F crosses μ_3, the periodic orbit shrinks and vanishes, and the low equilibrium undergoes a supercritical Hopf bifurcation. The low climate state has now become unstable: all trajectories starting near the origin spiral outwards and counter-clockwise until they finally reach the high climate state, as illustrated in Figure 3.3 for $F = -0.5$.

In summary, as we vary the climatological forcing, F, as a bifurcation parameter, the dynamical system (3.30) can have a unique steady state or two competing (stable) steady states, further distinguished by the presence or absence of an unstable periodic orbit. Three different climate scenarios occur: a single, less energetic, "low" climate; a single, more energetic, "high" climate; or a situation of competition between the two. The intermediate stage of competing climates further separates in two regimes, depending on the presence or absence of a periodic orbit. We choose the following four representative values for these distinct situations: $F = -0.1, -0.18, -0.25, -0.5$. Phase portraits for these different climatological forcing strengths are shown in Figure 3.3, while the coordinates of the corresponding stable equilibria are given in Table 3.1.

3.7. Additive and multiplicative stochastic noise and projected deterministic dynamics

To compare the original stochastic dynamics equations with their reduced counterpart in a systematic fashion, thereby evaluating the accuracy of the MTV procedure, we introduce two new parameters, θ and λ, which control in (3.13) and (3.21) the strength of the linear additive coupling and the nonlinear triad interactions, respectively. Thus, we consider

$$\frac{d\vec{x}}{dt} = D\vec{x} + (\omega + a_1x_1 + a_2x_2)\vec{x}^{\perp} + \vec{F} + \theta\vec{L}y + \lambda\begin{bmatrix} b_1x_2y \\ b_2x_1y \end{bmatrix}$$

$$\frac{dy}{dt} = -\theta\vec{L}\cdot\vec{x} + \lambda b_3x_1x_2 - \frac{\gamma}{\varepsilon}y + \frac{\sigma}{\sqrt{\varepsilon}}\dot{W}(t). \tag{3.31}$$

The reduced dynamics counterpart is given by (3.21), with $\vec{L}$ replaced by $\theta\vec{L}$ and b_i replaced by λb_i, $i = 1, 2, 3$.

In (3.31) all parameters, but ε, λ, θ, and $\vec{F} = [F, 0]^\top$, are fixed and set to the same values as in (3.29). Moreover, we choose

$$(3.32) \qquad b_1 = 0.25, \quad b_2 = 0.25, \quad b_3 = -0.5, \quad \vec{L} = \begin{bmatrix} c - 0.5 \\ 0.5 \end{bmatrix}, \quad \gamma = 1, \ \sigma = 1,$$

and set $F_3 = 0$ here for simplicity. We shall now systematically study the influence of the time scale ratio, ε, of the strength of the linear additive coupling, λ, of the strength of the nonlinear triad interaction, θ, and of the climatological forcing term, F. In particular, we wish to evaluate the relevance of the bifurcation sequence and the resulting four different climate scenarios, which we identified for the purely deterministic part of the dynamics in Section 3.6 in the full nonlinear system with noise.

Numerical method and statistics. The stochastic differential equations (3.13), (3.21), and (3.31) all fit in the general form

$$(3.33) \qquad d\vec{x} = f(\vec{x})\,dt + g(\vec{x})\,dW.$$

For the numerical solution of (3.33) we use a split-step strategy: the deterministic part is integrated with the standard fourth-order Runge–Kutta method, whereas the stochastic part is integrated with the Milstein scheme [**28**]. Hence the overall method converges strongly with order one. Since g does not depend on $\vec{x}$ in (3.13), the Milstein scheme reduces to the forward Euler method in that case. All computations were performed with a time step $\Delta t = 10^{-4}$, which was found to yield ample accuracy.

To generate the random numbers needed for the Brownian increments, we use the "Mersenne-Twister" algorithm by Matsumoto and Nishimura [**41**]. This algorithm produces a sequence of uniformly distributed pseudo-random numbers of period $2^{19937} - 1$. We transform the uniform distribution to a normal distribution by applying the standard Box–Muller transformation.

We shall compute the following statistics of the system: the auto-correlation, $\rho(s)$, the third order two-time moment, $K_3(s)$, and the fourth order two-time moment $K_4(s)$, given by

$$(3.34) \qquad \rho(s) = \frac{\langle (x(t+s) - \bar{x})(x(t) - \bar{x}) \rangle}{\langle (x(t) - \bar{x})^2 \rangle},$$

$$(3.35) \qquad K_3(s) = \frac{\langle (x(t+s) - \bar{x})^2 (x(t) - \bar{x}) \rangle}{\langle (x(t) - \bar{x})^2 \rangle^{3/2}},$$

$$(3.36) \qquad K_4(s) = \frac{\langle (x(t+s) - \bar{x})^2 (x(t) - \bar{x})^2 \rangle}{\langle (x(t) - \bar{x})^2 \rangle^2 + 2\langle (x(t+s) - \bar{x})(x(t) - \bar{x}) \rangle^2},$$

The correlation function $\rho(s)$ is a standard quantity of interest while the normalized higher moment correlations in (3.35), (3.36) provide more severe measures of non-Gaussianity [**17, 36**]. Here the terms inside "$\langle\ \rangle$" correspond to time averages and are always computed over $[0, T]$, $T = 10^5$; in the limit $T \to \infty$ they coincide with ensemble averages because of the underlying ergodic assumption. To avoid any transient behavior of initial stages, typically influenced by initial conditions etc., the statistics are only based on solution values $\vec{x}(t)$ past a spin-up

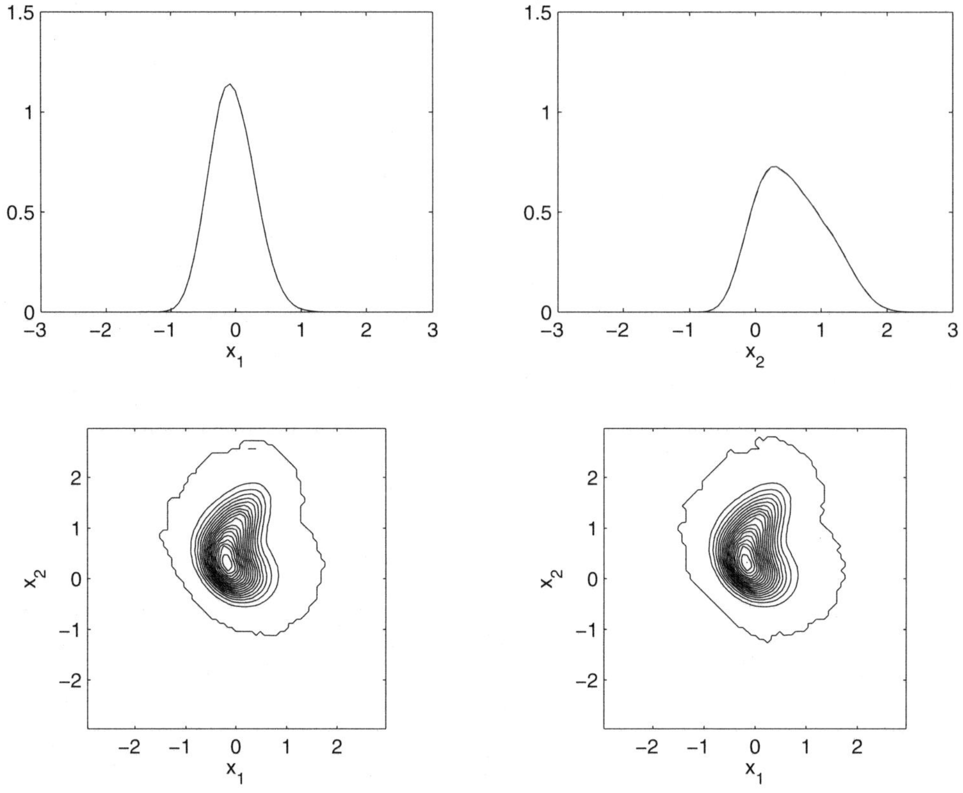

FIGURE 3.4. Linear additive coupling, low climate state ($F = -0.1$): comparison of the marginal and joint PDFs of x_1 and x_2 for $\theta = 0.5$, $\varepsilon = 0.1$. Top: the marginal PDF for the stochastic dynamics (plain) and for the reduced stochastic equations (dashed) are shown—here they coincide. Bottom: the joint PDF for the stochastic dynamics (left) and the reduced dynamics (right).

time. Here we always set that spin-up time to 10,000, that is to 10% of the entire simulation.

All terms in (3.34)–(3.36) involve the mean, $\bar{x} = \langle x(t) \rangle$, which is only known by the end of the computation. Thus a direct evaluation of (3.34)–(3.36) requires first computing and storing all the computed values, $\{x(t) | 0 \leq t \leq T\}$, then computing the mean, and finally evaluating the desired statistical quantity. Because of the sheer size of the data set, $\{x(t)\}$, which typically consists of 10^9 values, this approach is usually inefficient and slow. In the Appendix we show how to compute these statistical quantities together with the PDFs "on the fly", that is concurrently with $x(t)$, thereby avoiding the unnecessary storage of the entire realization $x(t)$.

Following the discussion in Section 3.6 we shall now systematically explore the parameter space for ε, λ, θ, F, first in the situation of pure linear additive coupling and then in the situation of nonlinear triad interactions.

Linear additive coupling: $\lambda = 0$. We begin with the weakest climatological forcing and set $F = -0.1$. For $\varepsilon = 0.1$ the time scales of the fast variable y and the slow variables $\vec{x}$ are only weakly correlated. When the strength of the linear coupling is small, that is $\theta = 0.1$, the joint probability distribution functions (PDFs) of x_1 and x_2 reduce to a narrow, essentially Gaussian distribution about the low climate state given in Table 3.1.

As we increase the interaction strength and set $\theta = 0.5$, the joint PDF, shown in Figure 3.4, adopts a broad kidney-shaped signature, while the mean of x_2 clearly shifts upwards. This shift is solely due to the linear interaction term and is perfectly captured by the reduced dynamics—see Figure 3.4. When the PDF of the full dynamics and that of the stochastic dynamics essentially coincide, the mean, variance, and the time correlations of the higher moments are usually hardly distinguishable; hence, we omit showing them in those instances.

Further increase in θ intensifies this trend, which is accurately predicted by the MTV-reduced stochastic dynamics for values of ε as large as $\varepsilon = 0.5$, even though this procedure used to derive the reduced stochastic dynamics (3.21) is rigorously valid only in the limit of vanishing ε. In the extreme situation of combined strong

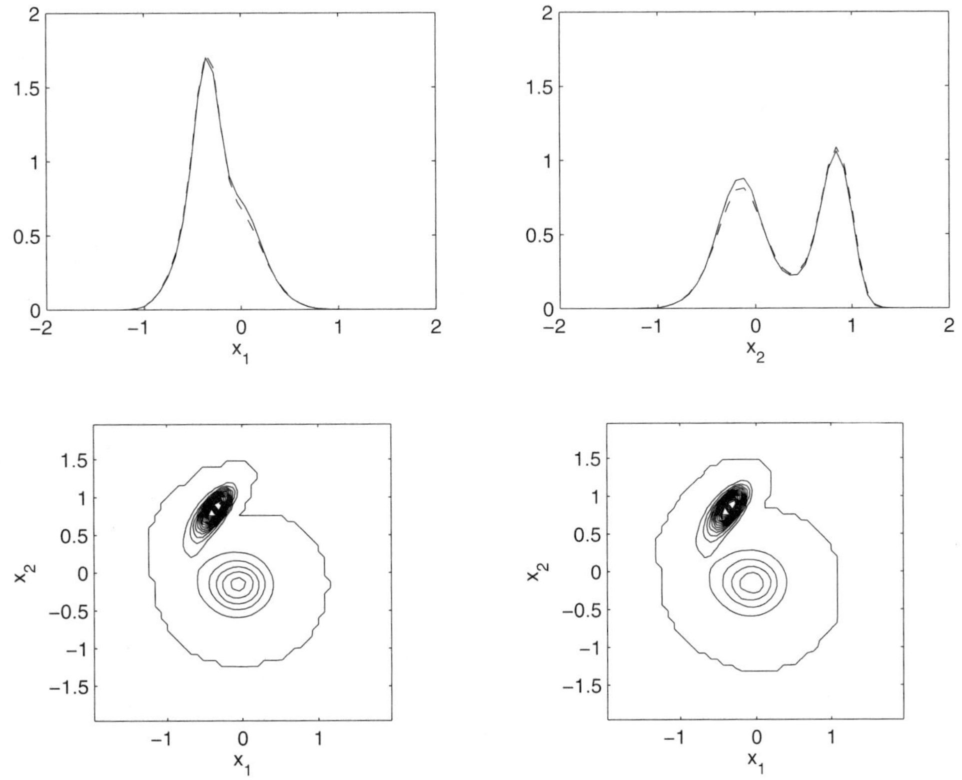

FIGURE 3.5. Linear additive coupling, competing climate states ($F = -0.25$): comparison of the marginal and joint PDFs of x_1 and x_2 for $\theta = 0.1$, $\varepsilon = 0.5$. Top: the marginal PDF for the stochastic dynamics (plain) and for the reduced stochastic equations (dashed) are shown. Bottom: the joint PDF for the stochastic dynamics (left) and the reduced dynamics (right).

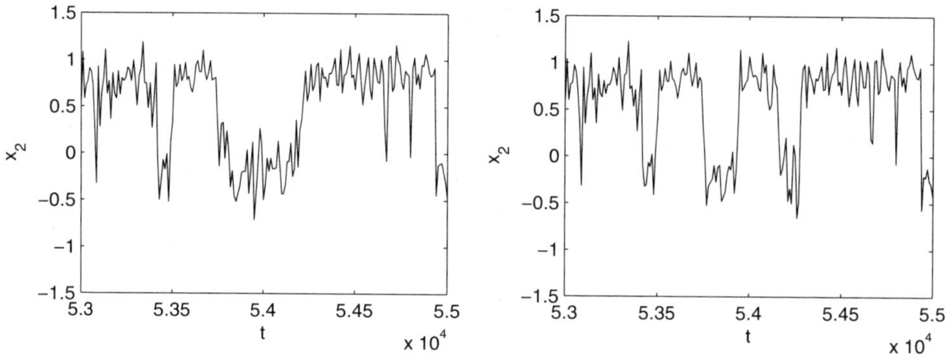

FIGURE 3.6. Linear additive coupling, competing climate states ($F = -0.25$): parts of the realizations $x_2(t)$ are shown for $\theta = 0.1$, $\varepsilon = 0.5$: full stochastic dynamics (left) and reduced dynamics (right).

linear coupling, $\theta = 1$, and little time scale separation, $\varepsilon = 0.5$, a small climate drift appears in the reduced dynamics, yet the error in the predicted mean remains below 10%.

As we increase the climatological forcing and cross the first bifurcation point of the deterministic dynamics located at $F = \mu_1$, the stochastic dynamics do not exhibit any significant qualitative change, unlike their bare truncation. In fact, for $F = -0.18$ the statistics still resemble our previous results with $F = -0.1$. Again, the reduced dynamics predict remarkably well the response to increases in the linear coupling or the noise level. This unexpected behavior, or rather the lack of it, shows that the bifurcation diagram of the bare truncation need not have any bearing on the fully stochastic regime.

However, as we further increase the climatological forcing to $F = -0.25$, two competing climate states indeed appear in the stochastic dynamics, as shown in Figure 3.5, for $\theta = 0.1$ and $\varepsilon = 0.5$. The locations of the two strong peaks in the double humped marginal PDF for x_2 and in the joint PDF coincide with the two low and climate states from the bare truncation in Table 3.1—see also Figure 3.3. To illustrate the actual random "hopping" between the two climate states we display part of the realization of x_2 in Figure 3.6. Although the left and right frames in Figure 3.6 clearly bear some resemblance—the same Brownian path $W(t)$ is used for their numerical integration—, differences are also evident. Smaller values of ε tend to widen the gaps in the time series of successive hopping events—see also [**18**]. In contrast, further increases in ε or θ tend to smear the distinct two climate states, while leading to a shifted kidney-shaped PDF similar to that shown in Figure 3.4.

In Figure 3.7, the time correlation functions of the various moments defined in (3.34)–(3.36) are shown. Again we observe the remarkable agreement between the statistics of the full dynamics and those of the reduced dynamics obtained from the MTV procedure. The large departure of K_4 from one underpins the non-Gaussianity of the underlying stochastic process.

As F crosses the final bifurcation point, the behavior of the stochastic dynamics parallels that of the deterministic part. Indeed for $F = -0.5$ and $\theta = 0.1$, the joint PDF appears narrow and centered about the high climate state, yet elongated in

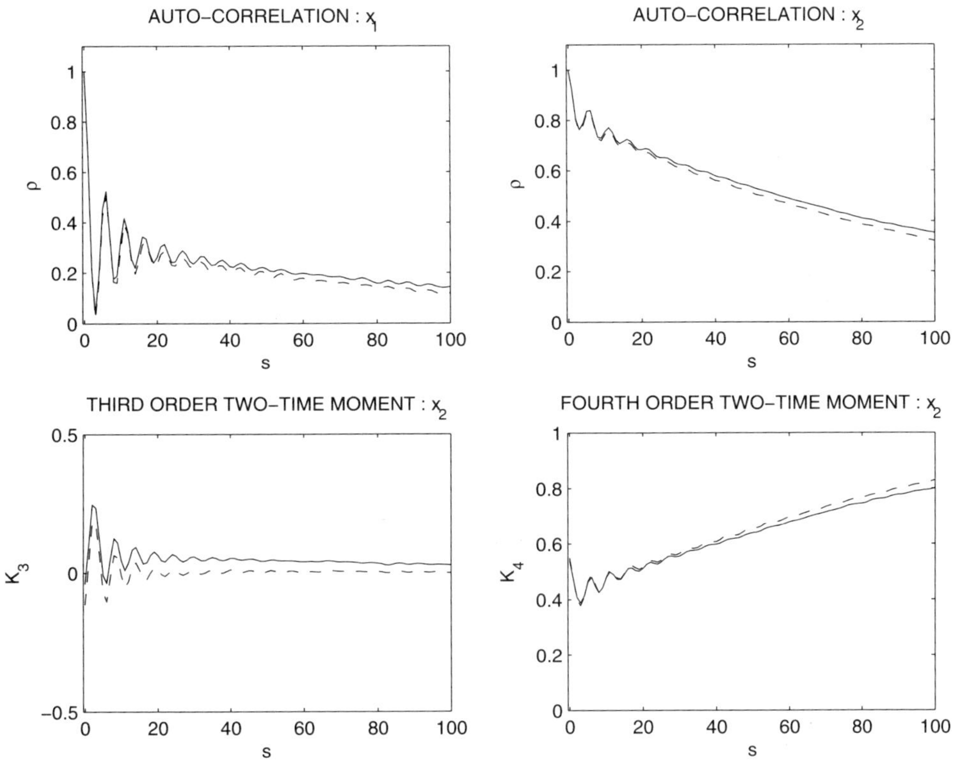

FIGURE 3.7. Linear additive coupling, competing climate states ($F = -0.25$). Statistics of the stochastic dynamics (plain) and the reduced dynamics (dashed) are shown for $\theta = 0.1$, $\varepsilon = 0.5$: auto-correlation $\rho(s)$ for x_1 (upper left); auto-correlation $\rho(s)$ for x_2 (upper right); third order two-time moment $K_3(s)$ for x_2 (lower left); fourth order two-time moment $K_4(s)$ for x_2 (lower right).

the direction of Ω (see Section 3.6); hence, the skewness is probably due to the underlying nonlinear dynamics in the vicinity of the (stable) equilibrium. Again, the reduced dynamics predict the long-time statistics of the slow variables $\vec{x}$ in the full stochastic system (3.30) with remarkable accuracy, even for ε as high as $\varepsilon = 0.5$, as shown in Figure 3.8. Here with $\varepsilon = 0.5$ and $\theta = 0.5$ we observe that the PDF flattens and curves downward, reaching even below the x_1-axis. To understand the nature of this peculiar behavior, we display in Figure 3.9 a short portion of the realization of $\vec{x}_2(t)$. The solution mainly hovers above $x_2 \geq 0.5$, attempting to reach the high climate state, while the added noise constantly perturbs its path creating wide excursions towards the lower, yet unstable, equilibrium—compare also with the lower right frame in Figure 3.3.

Nonlinear triad interactions: $\theta = 0$. We begin again with the weakest climatological forcing, $F = -0.1$, and consider (3.31) with $\theta = 0$. Even for strongly correlated time scales with $\varepsilon = 0.5$, the stochastic dynamics essentially remains tightly bound to the low climate state, with a highly peaked Gaussian PDF about it, as long as λ remains below 0.5. For $\lambda = 1$, however, the strong triad interactions

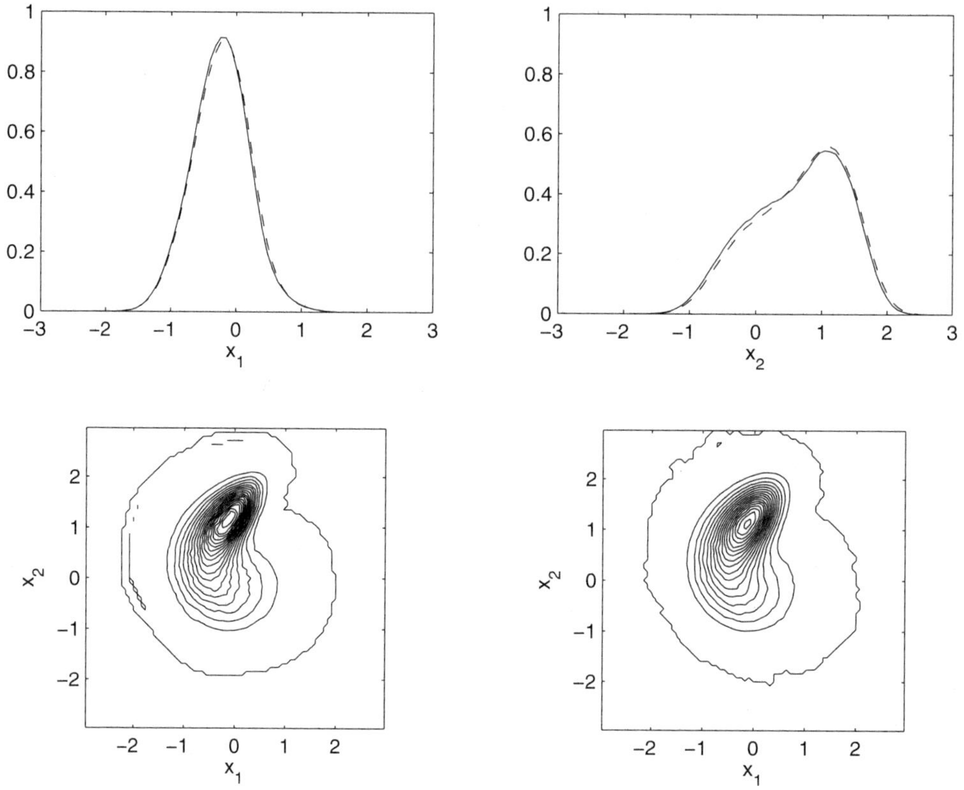

FIGURE 3.8. Linear additive coupling, high climate state ($F = -0.5$): comparison of the marginal and joint PDFs of x_1 and x_2 for $\theta = 0.5$, $\varepsilon = 0.5$. Top: the marginal PDF for the stochastic dynamics (plain) and for the reduced stochastic equations (dashed) are shown. Bottom: the joint PDF for the stochastic dynamics (left) and the reduced dynamics (right).

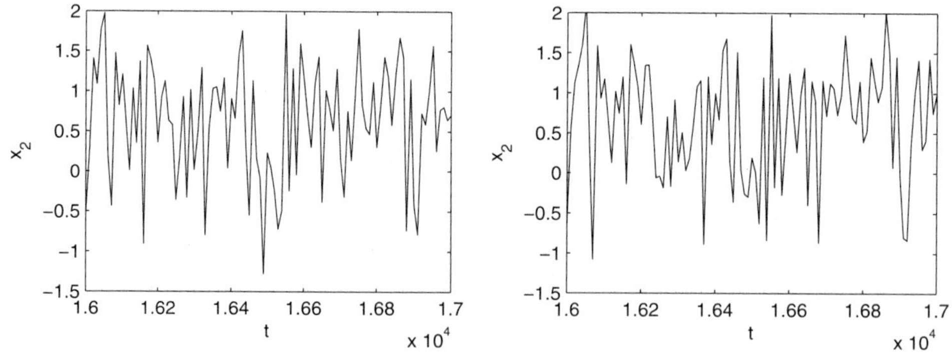

FIGURE 3.9. Linear additive coupling, high climate state ($F = -0.5$): parts of the realizations $x_2(t)$ are shown for $\theta = 0.5$, $\varepsilon = 0.5$. Full stochastic dynamics (left) and reduced dynamics (right).

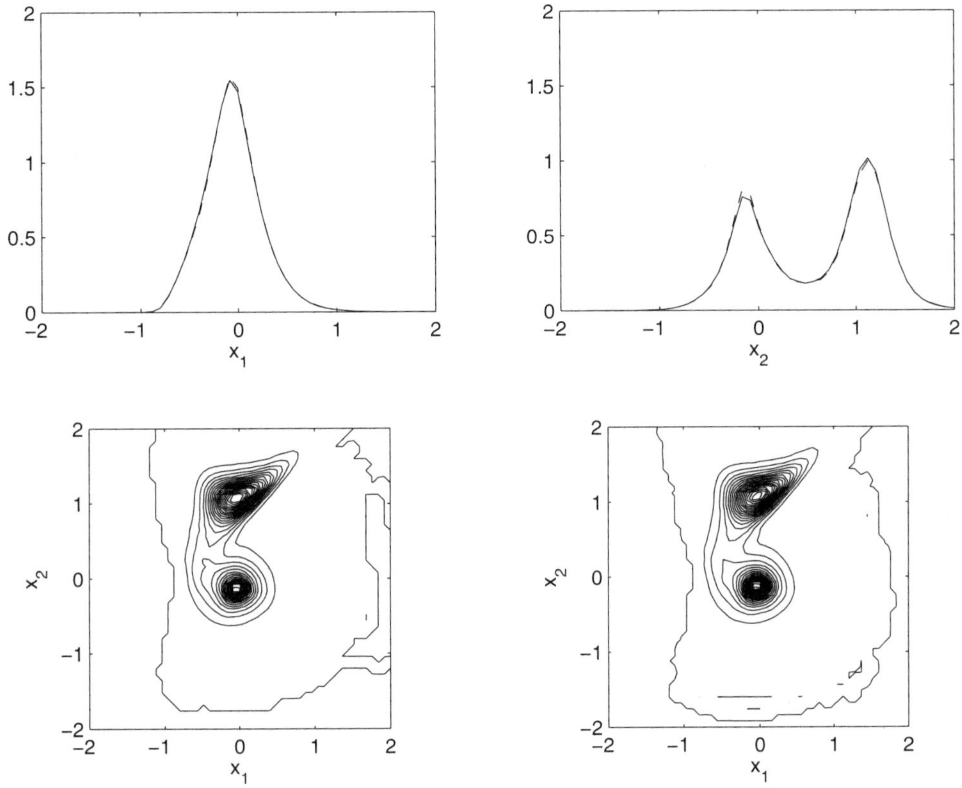

FIGURE 3.10. Nonlinear triad interactions, competing climate states ($F = -0.18$): comparison of the marginal and joint PDFs of x_1 and x_2 for $\lambda = 1$, $\varepsilon = 0.1$. Top: the marginal PDF for the stochastic dynamics (plain) and for the reduced stochastic equations (dashed) are shown – here they coincide. Bottom: the joint PDF for the stochastic dynamics (left) and the reduced dynamics (right).

induce intermittent far-reaching excursions above the region $x_2 \geq 1$. In all cases, the reduced dynamics obtained by the MTV procedure in the limit $\varepsilon \to 0$ capture remarkably well the full stochastic dynamics of the slow variables $\vec{x}$.

In contrast to the previous situation with additive linear coupling, the dynamics now sense the bifurcation of the bare truncation, as we set $F = -0.18$ and $\varepsilon = 0.1$. In Figure 3.10 we observe two well-defined competing climates. The seemingly random jumps between the upper and lower climate states, illustrated in Figure 3.11, demonstrates the incessant "hopping" of $\vec{x}(t)$ between the two underlying equilibria.

For $\varepsilon = 0.5$ the reduced dynamics capture the behavior of the full stochastic system with high fidelity until $\lambda = 0.5$. For $\lambda = 1$ and $\varepsilon = 0.5$, however, with combined large triad interactions and no time scale separation, the reduced dynamics reproduces only partially the PDFs of the full stochastic dynamics, as shown in Figure 3.12. Nonetheless, qualitatively the reduced dynamics still capture the main features of the statistics—see Figure 3.13. The large departure of $K_4(s)$ from one,

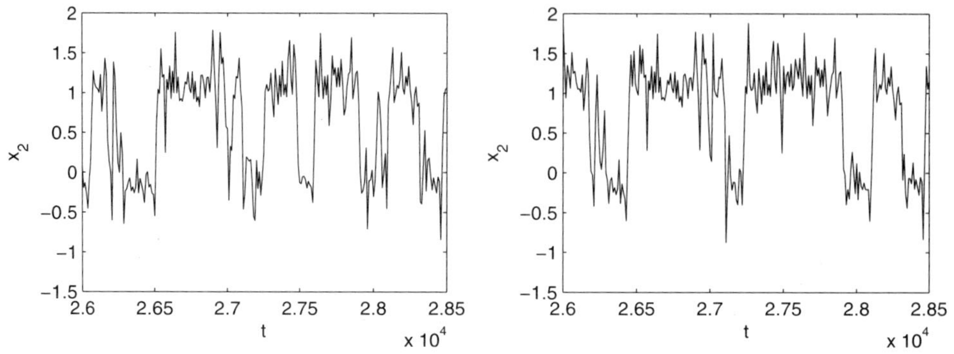

FIGURE 3.11. Nonlinear triad interactions, competing climate states ($F = -0.18$): parts of the realizations $x_2(t)$ are shown for $\lambda = 1$, $\varepsilon = 0.1$. Full stochastic dynamics (left) and reduced dynamics (right).

together with $K_3(s)$ close to zero, indicates the strong non-Gaussian nature of the underlying stochastic process.

As we further increase the strength of the climatological forcing and set $F = -0.25$, that is past the homoclinic bifurcation at μ_2, the PDFs tend to shift to higher values of x_2. Recall that here the low climate has a reduced basin of attraction bounded by the periodic orbit, which probably explains the apparent dominance of the high climate. Nevertheless a subtle interplay emerges between ε and λ, as the dominance sometimes shifts to the low climate for particular values of ε and λ.

To conclude our discussion we set $F = -0.5$ and consider a situation of no time scale separation and triad interactions with $\varepsilon = 0.5$ and $\lambda = 1$. Figure 3.14 displays the marginal PDFs of x_1 and x_2. Even in that extreme situation the reduced dynamics reproduce remarkably well the behavior of the full stochastic dynamics, centered about the high climate state, but for a slight climate drift below 5%. Like in the previous situation with linear additive coupling, the (unstable) low climate manifests its presence by inducing large intermittent deviations far below the x_1-axis. Again the time averaged statistics of the stochastic dynamics are well captured by the reduced equations, as shown in Figure 3.15.

3.8. Summary and discussion

The dynamical core of comprehensive general circulation models (GCM) for the atmosphere or ocean is governed essentially by quadratically nonlinear equations that involve a wide range of intrinsic time scales. Thus there is a great practical interest in deriving reduced models with many fewer degrees of freedom, which can capture the important long-time low frequency variability of such a system with reasonable accuracy. The MTV procedure proposed by Majda, Timofeyev, and Vanden–Eijnden [**34, 35, 36, 37**], is a systematic mathematical strategy which attempts to achieve this. It is based on the assumption that the degrees of freedom under consideration have been split into essential resolved modes, whose dynamics is the main objective, and unessential unresolved modes, which are to be eliminated in the reduced description.

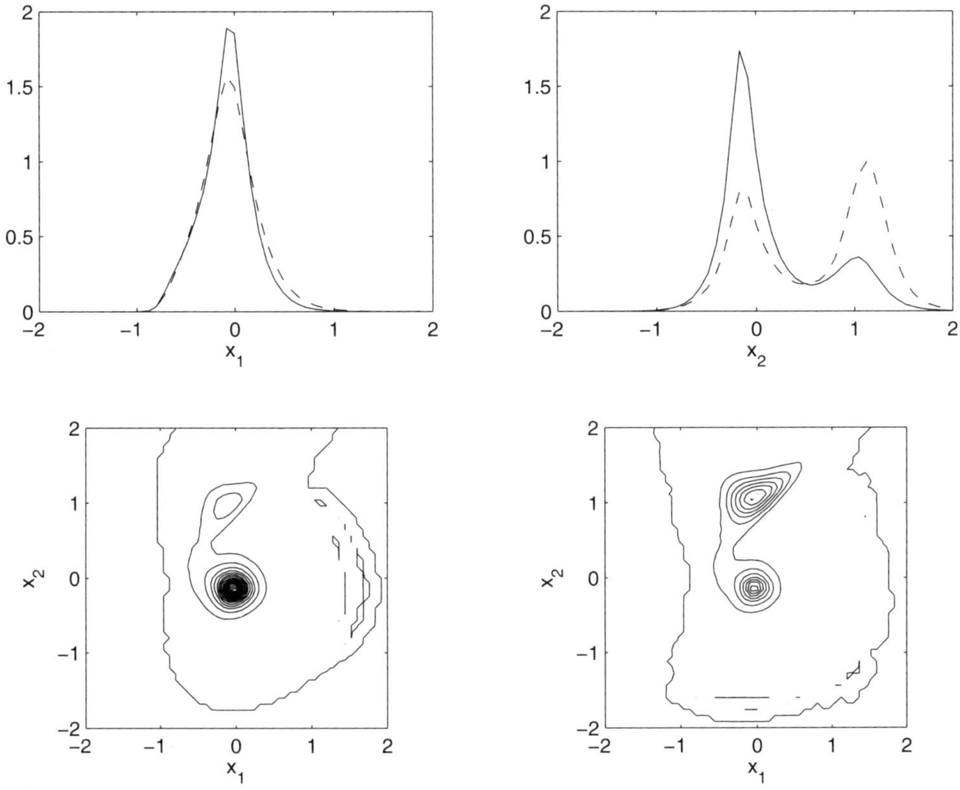

FIGURE 3.12. Nonlinear triad interactions, competing climate state ($F = -0.18$): comparison of the marginal and joint PDFs of x_1 and x_2 for $\lambda = 1$, $\varepsilon = 0.5$. Top: the marginal PDF for the stochastic dynamics (plain) and for the reduced stochastic equations (dashed) are shown. Bottom: the joint PDF for the stochastic dynamics (left) and the reduced dynamics (right).

In the first step of the MTV procedure, called *stochastic consistency*, the equations of motion for the unresolved modes are modified by representing the nonlinear self-interaction terms between unresolved modes by stochastic terms. In the second step of the procedure, called *stochastic mode elimination*, the equations of motion for the unresolved modes are then eliminated using projection techniques for stochastic differential equations. The elimination step is mathematically rigorous in the limit where the stochastic terms are infinitely fast, corresponding to the situation where the unresolved modes evolve much faster than the resolved ones.

Here we have considered the simplest version of a geophysical model with two (slow) climate variables and a single (fast) unresolved variable, with added stochastic forcing to represent the nonlinear interactions with the other unresolved modes. The time correlation between the fast and slow variables is measured by the single parameter, $\varepsilon > 0$. The MTV procedure, which is mathematically rigorous in the limit $\varepsilon \to 0$, then yields a system of *reduced stochastic differential equations* for the slow variables only, thereby effectively eliminating the fast unresolved mode. In

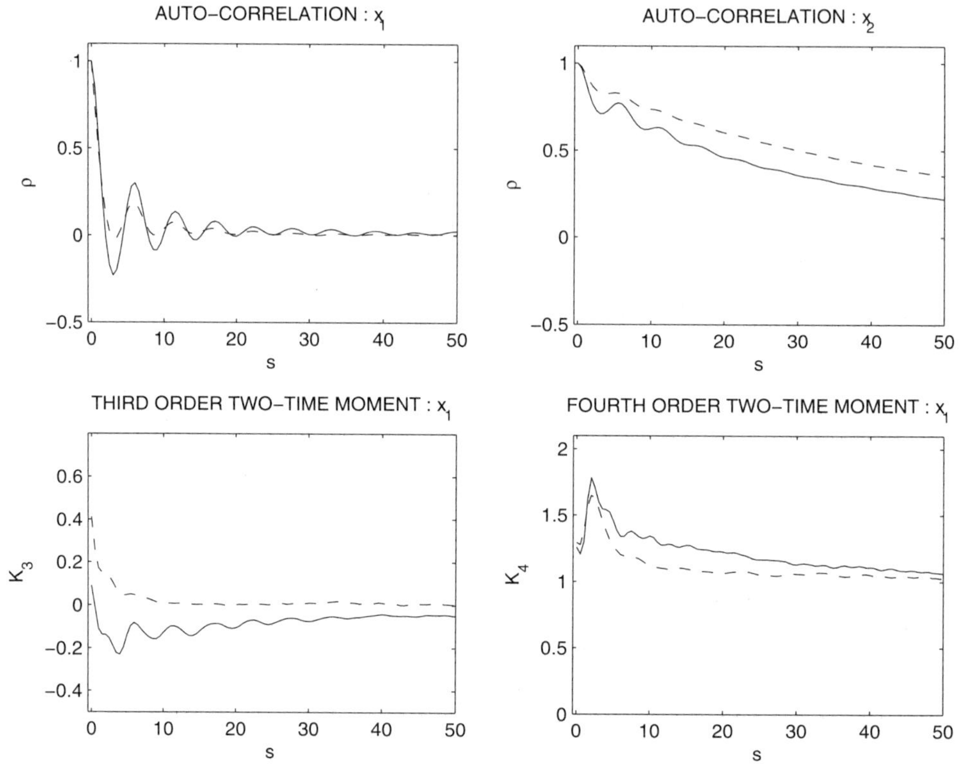

FIGURE 3.13. Nonlinear triad interactions, competing climate states ($F = -0.18$): statistics of the stochastic dynamics (plain) and the reduced dynamics (dashed) are shown for $\lambda = 1$, $\varepsilon = 0.5$: auto-correlation $\rho(s)$ for x_1 (upper left); auto-correlation $\rho(s)$ for x_2 (upper right); third order two-time moment $K_3(s)$ for x_1 (lower left); fourth order two-time moment $K_4(s)$ for x_1 (lower right).

addition to the deterministic dynamics term, the effective forcing now contains remarkable new features, such as augmented quadratic and cubic nonlinearity, altered mean forcing, and cross-diagonal dissipation, while the effective noise coefficient, contains both additive and multiplicative stochastic noise.

Although our model is elementary, it is a very instructive example for the competition between projected nonlinear dynamics and intrinsic stochastic noise. In particular, as the climatological forcing is varied, the purely deterministic underlying dynamics can have a unique steady state or two competing (stable) steady states, further distinguished by the presence or absence of an unstable periodic orbit. Hence, three different climate scenarios occur: a single, less energetic, "low" climate; a single, more energetic, "high" climate; or a situation of competition between the two. Moreover, the intermediate stage of competing climates further separates in two regimes, depending on the presence or absence of an unstable periodic orbit.

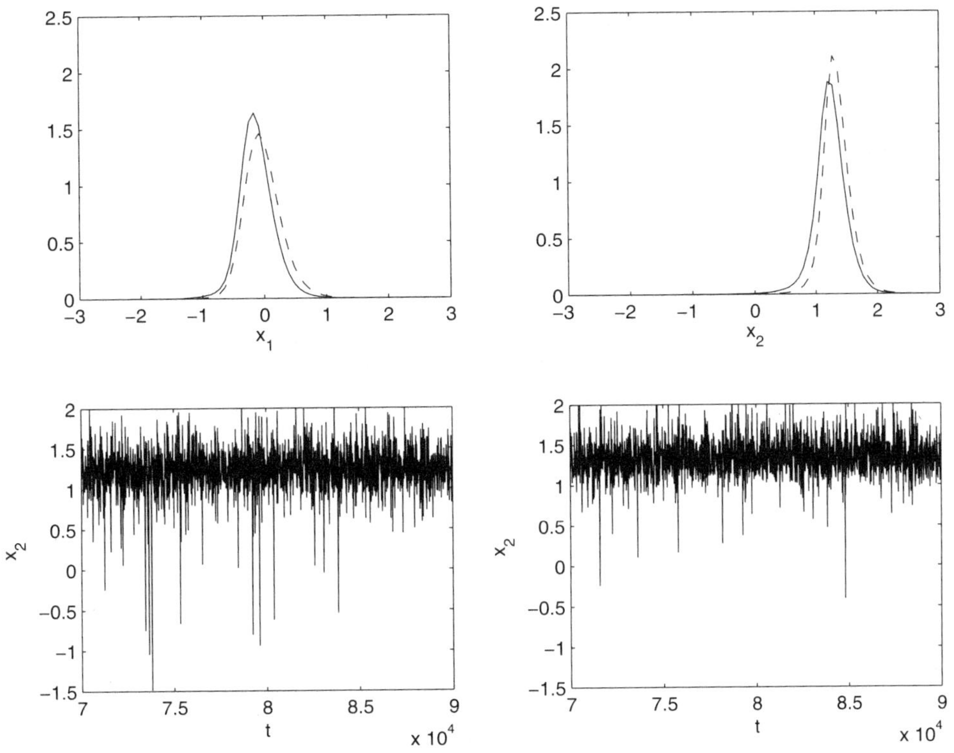

FIGURE 3.14. Nonlinear triad interactions, high climate state ($F = -0.5$). Top: the marginal PDF for the stochastic dynamics (plain) and for the reduced stochastic equations (dashed) are shown for $\lambda = 1$, $\varepsilon = 0.5$. Bottom: part of the realization $x_2(t)$ is shown for the stochastic dynamics (left) and the reduced dynamics (right).

Through a series of numerical experiments, the original three-dimensional stochastic dynamics are compared with the reduced system derived from the MTV procedure. Both the effects of added linear coupling and nonlinear triad interactions are systematically tested. Typically for values of $\varepsilon \leq 0.5$ the reduced stochastic dynamics predict remarkably well the long-time statistics of the original system; often with such high precision, that the probability density functions essentially coincide. For higher values of ε, we sometimes observe small climate drifts in the prediction within a 10% range of the mean. Nevertheless, the long-time statistics, such as the auto-correlation and the two-time higher moments, are usually predicted with high fidelity. Although the qualitative behavior of the underlying purely deterministic dynamics can predict the behavior of the stochastic regime in some instances, we find many situations where the effective forcing due to the noise dramatically changes the qualitative behavior of the full stochastic regime, so that the deterministic bifurcation diagram becomes irrelevant. Yet even in those instances, the emerging behavior is well captured by the reduced stochastic equations obtained from the MTV procedure.

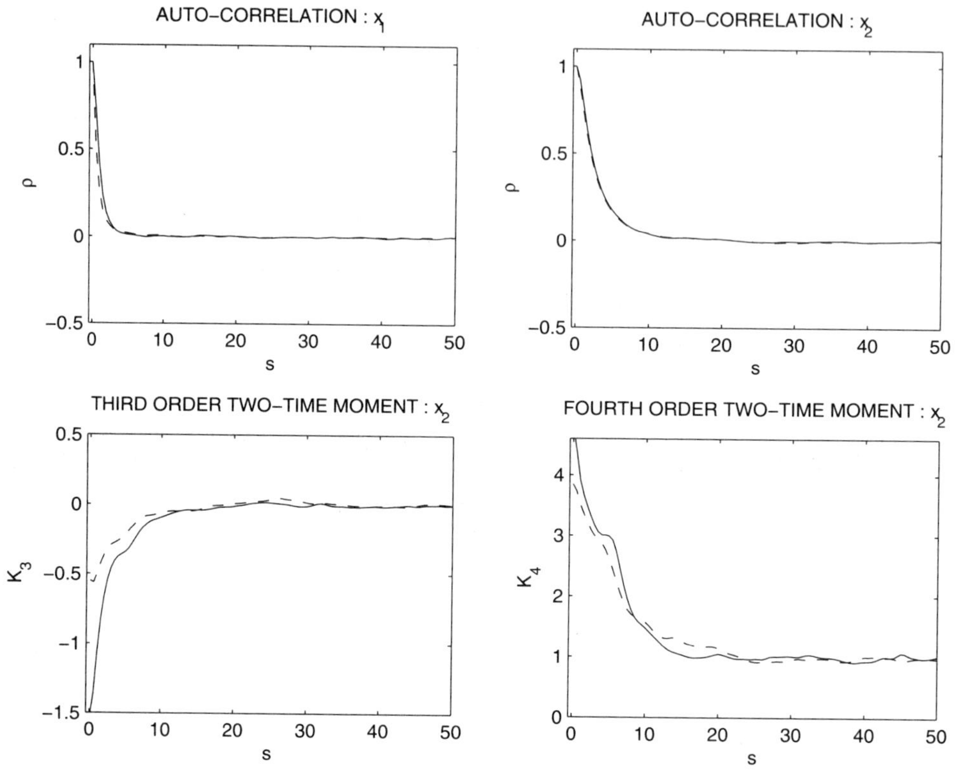

FIGURE 3.15. Nonlinear triad interactions, high climate state ($F = -0.5$): statistics of the stochastic dynamics (plain) and the reduced dynamics (dashed) are shown for $\lambda = 1$, $\varepsilon = 0.5$: auto-correlation $\rho(s)$ for x_1 (upper left); auto-correlation $\rho(s)$ for x_2 (upper right); third order two-time moment $K_3(s)$ for x_2 (lower left); fourth order two-time moment $K_4(s)$ for x_2 (lower right).

Appendix

When considering stochastic dynamical systems, one is usually not interested in any particular realization *per se*, but instead in various statistical quantities, such as the auto-correlation $\rho(s)$, or the high-order two-time moments $K_3(s)$ and $K_4(s)$, defined previously in (3.34)–(3.36). All these quantities involve the mean, $\bar{x}$ or $\langle x(t) \rangle$, which in fact is only known by the end of the computation. Thus a direct evaluation of (3.34)–(3.36) typically requires first computing and storing all the computed values, $x(t)$, then computing the mean, and finally evaluating the desired statistical quantity. Because of the sheer size of the data set, $x(t)$, this approach involves many read/write operations on mass storage units, such as a hard disk, which is slow and inefficient. Here we show how to compute these statistical quantities together with the PDFs "on the fly", that is concurrently with $x(t)$. Thus, we avoid the unnecessary storage of the entire realization $\{x(t) | 0 \leq t \leq T\}$.

Statistics. To compute the statistics "on the fly", we first expand all terms in (3.34) to obtain

$$\rho(s) = \frac{\langle x(t+s)x(t)\rangle - \langle x(t+s)\bar{x}\rangle - \langle \bar{x}x(t)\rangle + \bar{x}^2}{\langle x^2(t)\rangle - 2\langle \bar{x}x(t)\rangle + \bar{x}^2}. \tag{3.37}$$

Since $\bar{x}$ is constant and $\langle x(t)\rangle = \langle x(t+s)\rangle$ for any s, we can further simplify (3.37), which yields

$$\rho(s) = \frac{\langle x(t+s)x(t)\rangle - \bar{x}^2}{\langle x^2(t)\rangle - \bar{x}^2\rangle}. \tag{3.38}$$

Similarly, we find

$$K_3(s) = \frac{\langle x^2(t+s)x(t)\rangle - \bar{x}\langle x^2(t)\rangle - 2\bar{x}\langle x(t+s)x(t)\rangle + 2\bar{x}^3}{(\langle x^2(t)\rangle - \bar{x}^2)^{3/2}}, \tag{3.39}$$

and

$$\begin{aligned} K_4(s) = {} & [\langle x^2(t+s)x^2(t)\rangle - 2\bar{x}\langle x^2(t+s)x(t)\rangle - 2\bar{x}\langle x(t+s)x^2(t)\rangle \\ & + 2\bar{x}^2\langle x^2(t)\rangle + 4\bar{x}^2\langle x(t+s)x(t)\rangle - 3\bar{x}^4] \\ & /[(\langle x^2(t)\rangle - \bar{x}^2)^2 + 2(\langle x(t+s)x(t)\rangle - \bar{x}^2)^2]. \end{aligned} \tag{3.40}$$

To compute $\rho(s)$, $K_3(s)$, and $K_4(s)$ it is therefore sufficient to compute the moments $\langle x^k(t+s)x^\ell(t)\rangle$, with $k,\ \ell = 1, 2$. These four moments are computed for discrete values of $s \in [0, s_{\max}]$ by summation "on the fly", that is by summation over all time steps concurrently with the computation of $x(t)$. Finally their long-time averages are used in (3.38)–(3.40) to evaluate the various statistical quantities of interest for each s. Thus, the values of $x(t)$ only need to be stored over the interval $[t, t + s_{\max}]$.

Probability density functions. The probability density functions (PDFs) are usually computed by standard "bin counting". Here we let the interval $[a, b]$ denote the numerical range of the (discrete approximation to the) stochastic process $x(t)$. Next, we decompose $[a, b]$ in a fixed number of sub-intervals of equal length, called "bins", typically 50 or 100 bins. During the realization of $\vec{x}(t)$, the number of instances the value of $\vec{x}$ occurs inside any sub-interval is recorded for every bin. Finally, that number is divided by the total number of time steps, which yields a genuine PDF normalized to one over $[a, b]$. Clearly the same procedure can be used for joint PDFs in two dimensions.

When computing the PDF "on the fly", again to avoid the inefficient storage of the entire realization, the range of $\{x(t_i) \mid 0 \le t_i \le T\}$ is usually unknown a priori, so that some percentage of the values will lie outside $[a, b]$. That percentage then needs to be taken into account in the final normalization over the entire x-axis, to ensure that the PDF is properly weighted over $[a, b]$.

References

1. U. Achatz and G. Branstator, *A two-layer model with empirical linear corrections and reduced order for studies of internal climate variability*, J. Atmospheric Sci. **56** (1999), 3140–3160.
2. L. Arnold, *Stochastic differential equations*, Wiley-Interscience, New York, 1974.
3. C. Beck and F. Schlögl, *Thermodynamics of chaotic systems*, Cambridge Nonlinear Science Series, vol. 4, Cambridge University Press, New York, 1997.
4. G. Branstator, *Organization of storm track anomalies by recurring low-frequency circulation anomalies*, J. Atmospheric Sci. **52** (1995), 207–226.

5. G. Branstator and S. E. Haupt, *An empirical model of barotropic atmospheric dynamics and its response to tropical forcing*, Journal of Climate **11** (1998), 2645–2667.
6. J. G. Charney and J. G. DeVore, *Multiple flow equilibria in the atmosphere and blocking*, J. Atmospheric Sci. **36** (1979), 1205–1216.
7. D. T. Crommelin, *Regime transitions and heteroclinic connections in a barotropic atmosphere*, J. Atmospheric Sci. **60** (2003), 229–246.
8. D. T. Crommelin and A. J. Majda, *Strategies for model reduction: Comparing different optimal bases*, J. Atmospheric Sci. **61** (2004), 2206–2217.
9. T. DelSole, *Optimally persistent patterns in time-varying fields*, J. Atmospheric Sci. **58** (2001), 1341–1356.
10. ———, *Stochastic models of quasigeostrophic turbulence*, Surveys in Geophysics (2004), in press.
11. T. DelSole and B. F. Farrel, *Quasi-linear equilibration of a thermally maintained stochastically excited jet in a quasigeostrophic model*, J. Atmospheric Sci. **53** (1996), 1781–1797.
12. H. E. DeSwart and J. Grasman, *Effect of stochastic perturbations on a low-order spectral model of the atmospheric circulation*, Tellus **39A** (1987), 10–24.
13. E. Doedel and X. J. Wang, AUTO94: *Software for continuation and bifurcation problems in ordinary differential equations*, Tech. Report CRPC-95-2, Caltech, 1995.
14. J. Egger, *A stochastic model for the angular momentum budget of latitude belts*, in press.
15. ———, *Master equations for climate parameter sets*, Climate Dynamics **18** (2001), 169–177.
16. J. Egger and K.-P. Hoinka, *Covariance analysis of the global atmospheric axial angular momentum budget*, preprint, 2002.
17. Ch. Franzke, A. J. Majda, and E. Vanden Eijnden, *Low-order stochastic mode reduction for a realistic barotropic model climate*, J. Atmospheric Sci., in press.
18. M. I. Freidlin and A. D. Wentzell (eds.), *Random perturbations of dynamical systems*, Springer-Verlag, New-York, 1984.
19. C. W. Gardiner, *Handbook of Stochastic Methods*, Springer-Verlag, Berlin, 1985.
20. M. Ghil and S. Childress, *Topics in geophysical fluid dynamics, atmospheric dynamics, and climate dynamics*, Springer-Verlag, New York, 1987.
21. M. Ghil and A. W. Robertson, *"Waves" vs. "Particles" in the atmosphere's phases space: A pathway to long-range forecasting?*, Proc. Nat. Acad. Sci. USA **99** (2002), 2493–2500.
22. S. M. Griffies and K. Bryan, *A predictability study of simulated North Atlantic multidecadal variability*, Climate Dynamics **13** (1997), 459–487.
23. K. Hasselmann, *Stochastic climate models. I. theory*, Tellus **28** (1976), 473–485.
24. H. Itoh and M. Kimoto, *Multiple attractors and chaotic itinerancy in a quasi-geostrophic model with realistic topography: Implications for weather regimes and low-frequency variability*, J. Atmospheric Sci. **53** (1996), 2217–2231.
25. F.-F. Jin and M. Ghil, *Intraseasonal oscillations in the extratropics: Hopf bifurcation and topographic instabilities*, J. Atmospheric Sci. **47** (1990), 3007–3022.
26. R. Z. Khasminsky, *Principle of averaging for parabolic and elliptic differential equations and for Markov processes with small diffusion*, Theory Probab. Appl. **8** (1963), 1–21.
27. M. Kimoto and M. Ghil, *Multiple flow regimes in the Northern Hemisphere winter. II. Sectorial regimes and preferred transitions*, J. Atmospheric Sci. **50** (1993), 2645–2673.
28. P. E. Kloeden and E. Platen, *Numerical solution of stochastic differential equations*, Springer-Verlag, New York, 1995.
29. T. G. Kurtz, *A limit theorem for perturbed operators semigroups with applications to random evolution*, J. Funct. Anal. **12** (1973), 55–67.
30. F. Kwasniok, *The reduction of complex dynamical systems using principal interaction patterns*, Phys. D **92** (1996), 28–60.
31. ———, *Empirical low-order models of barotropic flow*, J. Atmospheric Sci. **61** (2004), 235–245.
32. B. Legras and M. Ghil, *Persistent anomalies, blocking and variations in atmospheric predictability*, J. Atmospheric Sci. **42** (1985), 433–471.
33. A. J. Majda and I. I. Timofeyev, *Low-dimensional choatic dynamics versus intrinsic stochastic noise: A paradigm model*, Phys. D **199** (2004), 339–368.
34. A. J. Majda, I. Timofeyev, and E. Vanden-Eijnden, *Models for stochastic climate prediction*, Proc. Nat. Acad. Sci. **96** (1999), 14687–14691.

35. ______, *A mathematical framework for stochastic climate models*, Comm. Pure Appl. Math. **54** (2001), 891–974.
36. ______, *A priori tests of a stochastic mode reduction strategy*, Phys. D **170** (2002), 206–252.
37. ______, *Systematic strategies for stochastic mode reduction in climate*, J. Atmospheric Sci. **60** (2003), 1705–1722.
38. A. J. Majda and X. Wang, *Nonlinear dynamics and statistical theories for basic geophysical flows*, Cambridge University Press, 2005.
39. S. L. Marcus, M. Ghil, and J. O. Dickey, *The extratropical 40-day oscillation in the* UCLA *general circulation model*. I. *Atmospheric angular momentum*, J. Atmospheric Sci. **51** (1994), 1431–1446.
40. ______, *The extratropical 40-day oscillation in the UCLA general circulation model*. II. *Spatial structure*, J. Atmospheric Sci. **53** (1996), 1993–2014.
41. M. Matsumoto and T. Nishimura, *Mersenne twister: A 623-dimensionally equidistributed uniform pseudorandom number generator*, ACM TOMACS **8** (1998), 3–30.
42. M. Newman, P. Sardeshmukh, and C. Penland, *Stochastic forcing of the wintertime extratropical flow*, J. Atmospheric Sci. **54** (1997), 435–455.
43. F. M. Selten, *A statistical closure of a low-order barotropic model*, J. Atmospheric Sci. **54** (1997), 1085–1093.
44. K. E. Trenberth, G. Branstator, D. Karoly, A. Kumar, N.-C. N.-C. Lau, and C. Ropelewski, *Progress during TOGA in understanding and modeling global teleconnections associated with tropical sea surface temperature*, J. Geophys. Res. (1998), 14291–14324.
45. R. Vautard and B. Legras, *On the source of midlatitude low-frequency variability*. II. *Nonlinear equilibration of weather regimes*, J. Atmospheric Sci. **45** (1988), 2845–2867.
46. K. M. Weickmann, W. Robinson, and C. Penland, *Stochastic and oscillatory forcing of global atmospheric angular momentum*, J. Geophys. Res. **105** (2000), 15543–15557.
47. J. A. Whitaker and P. D. Sardeshmukh, *A linear theory of extratropical synoptic eddy statistics*, J. Atmospheric Sci. **55** (1998), 237–258.

Titles in This Series

25 **Andrew J. Majda, Rafail V. Abramov, and Marcus J. Grote,** Information theory and stochastics for multiscale nonlinear systems, 2005

24 **Dana Schlomiuk, Andreĭ A. Bolibrukh, Sergei Yakovenko, Vadim Kaloshin, and Alexandru Buium,** On finiteness in differential equations and Diophantine geometry, 2005

23 **J. J. M. M. Rutten, Marta Kwiatkowska, Gethin Norman, and David Parker,** Mathematical techniques for analyzing concurrent and probabilistic systems, 2004

22 **Montserrat Alsina and Pilar Bayer,** Quaternion orders, quadratic forms, and Shimura curves, 2004

21 **Andrei Tyurin,** Quantization, classical and quantum field theory and theta functions, 2003

20 **Joel Feldman, Horst Knörrer, and Eugene Trubowitz,** Riemann surfaces of infinite genus, 2003

19 **L. Lafforgue,** Chirurgie des grassmanniennes, 2003

18 **G. Lusztig,** Hecke algebras with unequal parameters, 2003

17 **Michael Barr,** Acyclic models, 2002

16 **Joel Feldman, Horst Knörrer, and Eugene Trubowitz,** Fermionic functional integrals and the renormalization group, 2002

15 **José I. Burgos Gil,** The regulators of Beilinson and Borel, 2002

14 **Eyal Z. Goren,** Lectures on Hilbert modular varieties and modular forms, 2002

13 **Michael Baake and Robert V. Moody, Editors,** Directions in mathematical quasicrystals, 2000

12 **Masayoshi Miyanishi,** Open algebraic surfaces, 2001

11 **Spencer J. Bloch,** Higher regulators, algebraic K-theory, and zeta functions of elliptic curves, 2000

10 **James D. Lewis,** A survey of the Hodge conjecture, Second Edition, 1999

9 **Yves Meyer,** Wavelets, vibrations and scaling, 1998

8 **Ioannis Karatzas,** Lectures on the mathematics of finance, 1996

7 **John Milton,** Dynamics of small neural populations, 1996

6 **Eugene B. Dynkin,** An introduction to branching measure-valued processes, 1994

5 **Andrew Bruckner,** Differentiation of real functions, 1994

4 **David Ruelle,** Dynamical zeta functions for piecewise monotone maps of the interval, 1994

3 **V. Kumar Murty,** Introduction to Abelian varieties, 1993

2 **M. Ya. Antimirov, A. A. Kolyshkin, and Rémi Vaillancourt,** Applied integral transforms, 1993

1 **D. V. Voiculescu, K. J. Dykema, and A. Nica,** Free random variables, 1992